REPORT OF A WORKSHOP
on PREDICTABILITY &
LIMITS-TO-PREDICTION
in HYDROLOGIC SYSTEMS

Committee on Hydrologic Science

Water Science and Technology Board
Board on Atmospheric Sciences and Climate
Division on Earth and Life Studies

National Research Council

NATIONAL ACADEMY PRESS
Washington, D.C.

NATIONAL ACADEMY PRESS • 2101 Constitution Avenue, N.W. • Washington, DC 20418

NOTICE: The project that is the subject of this report was approved by the Governing Board of the National Research Council, whose members are drawn from the councils of the National Academy of Sciences, the National Academy of Engineering, and the Institute of Medicine. The members of the committee responsible for the report were chosen for their special competences and with regard for appropriate balance.

Support for this project was provided by the Army Research Office, National Aeronautics and Space Administration under Grant No. NAG5-8651, National Oceanic and Atmospheric Administration, National Science Foundation under Grant No. EAR-9986796, National Weather Service, U.S. Environmental Protection Agency under Grant No. X-2895301, and U.S. Geological Survey. Any opinions, findings, and conclusions or recommendations expressed in this material are those of the authors and do not necessarily reflect the views of the sponsors.

International Standard Book Number 0-309-08347-8

Predictability and Limits-To-Prediction in Hydrologic Systems is available from the National Academy Press, 2101 Constitution Avenue, N.W., Washington, DC 20418, (800) 624-6242 or (202) 334-3313 (in the Washington metropolitan area); internet <http://www.nap.edu>.

Cover photo, Great Falls, Virginia, courtesy of Sharon Gervasoni.

Printed in the United States of America

THE NATIONAL ACADEMIES
Advisers to the Nation on Science, Engineering, and Medicine

National Academy of Sciences
National Academy of Engineering
Institute of Medicine
National Research Council

The **National Academy of Sciences** is a private, nonprofit, self-perpetuating society of distinguished scholars engaged in scientific and engineering research, dedicated to the furtherance of science and technology and to their use for the general welfare. Upon the authority of the charter granted to it by the Congress in 1863, the Academy has a mandate that requires it to advise the federal government on scientific and technical matters. Dr. Bruce M. Alberts is president of the National Academy of Sciences.

The **National Academy of Engineering** was established in 1964, under the charter of the National Academy of Sciences, as a parallel organization of outstanding engineers. It is autonomous in its administration and in the selection of its members, sharing with the National Academy of Sciences the responsibility for advising the federal government. The National Academy of Engineering also sponsors engineering programs aimed at meeting national needs, encourages education and research, and recognizes the superior achievement of engineers. Dr. Wm. A. Wulf is president of the National Academy of Engineering.

The **Institute of Medicine** was established in 1970 by the National Academy of Sciences to secure the services of eminent members of appropriate professions in the examination of policy matters pertaining to the health of the public. The Institute acts under the responsibility given to the National Academy of Sciences by its congressional charter to be an adviser to the federal government and, upon its own initiative, to identify issues of medical care, research, and education. Dr. Kenneth I. Shine is president of the Institute of Medicine.

The **National Research Council** was organized by the National Academy of Sciences in 1916 to associate the broad community of science and technology with the Academy's purposes of furthering knowledge and advising the federal government. Functioning in accordance with general policies determined by the Academy, the Council has become the principal operating agency of both the National Academy of Sciences and the National Academy of Engineering in providing services to the government, the public, and the scientific and engineering communities. The Council is administered jointly by both Academies and the Institute of Medicine. Dr. Bruce M. Alberts and Dr. Wm A. Wulf are chairman and vice chairman, respectively, of the National Research Council.

COMMITTEE ON HYDROLOGIC SCIENCE[1]

DARA ENTEKHABI[2], *Chair*, Massachusetts Institute of Technology, Cambridge
MARY P. ANDERSON, University of Wisconsin, Madison
RONI AVISSAR, Rutgers University, New Brunswick, New Jersey
ROGER C. BALES, University of Arizona, Tucson
GEORGE M. HORNBERGER, University of Virginia, Charlottesville
WILLIAM K. NUTTLE, Consultant, West Palm Beach, Florida
MARC B. PARLANGE, Johns Hopkins University, Baltimore, Maryland
CHRISTA PETERS-LIDARD, Georgia Institute of Technology, Atlanta (through September 2001)
KENNETH W. POTTER, University of Wisconsin, Madison
JOHN O. ROADS, Scripps Institution of Oceanography, LaJolla, California
JOHN L. WILSON, New Mexico Tech, Socorro
ERIC F. WOOD, Princeton University, New Jersey

Staff

WILLIAM S. LOGAN, Staff Officer
VAUGHAN TUREKIAN[2], Staff Officer
ANITA A. HALL, Project Assistant

Editor

RHONDA BITTERLI

---

[1] The activities of the Committee on hydrologic Science are overseen and supported by the NRC's Water Science and Technology Board (lead) and Board on Atmospheric Sciences and Climate (see Appendix C).
[2] Dara Entekhabi and Vaughan Turekian were the chair and responsible staff officer, respectively, for the predictability workshop.

Preface

The Committee on Hydrologic Science (COHS) of the National Research Council (NRC) is engaged in studying the priorities and future strategies for hydrologic science. In order to involve a broad community representation, COHS is organizing workshops on priority topics in hydrologic science. These efforts will culminate in reports from the NRC on the individual workshops as well as a synthesis report on strategic directions in hydrologic science. The first workshop—Predictability and Limits-to-Prediction in Hydrologic Systems—was held at the National Center for Atmospheric Research in Boulder, Colorado, September 21–22, 2000. Fourteen technical presentations covered basic research and understanding, model formulations and behavior, observing strategies, and transition to operational predictions (see Appendix A for the agenda).

Understanding the limits of prediction directly impacts the activities and mission goals of federal, state, and local agencies, the relevance of research in the academic community, the engineering practices in industry, and the safety and awareness of the public at large when it comes to water issues. In many contexts (be it predicting the dispersion of a subsurface contaminant plume, the forecasting of a flash flood, the evolution of ecohydrologic systems, or the seasonal variations in stream water chemistry), data and models are used to gain insight to an event. The event may be in the future (classical forecasting), it may simply be downstream, or it may be the outcome of a particular perturbation of the system.

In all these cases, predictions should require (1) fundamental understanding of the dynamics of the system and propagation of perturbations or uncertainty through it, (2) adequate data to characterize the system states, and (3) procedures for producing the expected evolution of the system including interactions among its components. A prediction system additionally requires mechanisms for measuring the accuracy of predictions (often referred to as forecast skill), communicating the predictions to users in an effective manner, and using feedback from the users and the system's performance to improve the prediction and prediction delivery systems. This brief report is directed to understanding the common features associated with these issues in hydrologic science.

The premise of the workshop was that meaningful and robust prediction systems are built on basic understanding of predictability. Predictability research is directed toward understanding (1) remote and local sources of variability, (2) propagation of uncertainty and variability in hydrologic systems, (3) strategies for identifying and characterizing memories, information pathways, and feedback in hydrologic systems, and (4) quantifying intrinsic and model-derived limits to prediction.

Preceding the workshop the participants were provided with a white paper on the topic that was prepared by COHS members. The white paper was designed to provoke thinking in several critical directions among the participants. In the paper, six main questions were introduced. These questions reemerged during the workshop in order to initiate discussions during three panel-discussion periods. The questions are:

1. Are there predictable aspects of terrestrial hydrology that can enhance atmospheric weather and climate predictability?

2. What are the stability and feedback characteristics of two-way coupled subsurface, surface, and atmospheric hydrologic systems? How do they impact predictability?

3. What are the conceptual and model frameworks required to define limits-to-prediction in hydrologic systems?

4. What are the data and records requirements for estimating the inherent limits-to-prediction directly from observations?

5. What are the opportunities for extending the lead time and accuracy of hydrologic predictions based on predictable weather and climate patterns so that the predictions meet the requirements of water resource and other applications?

6. What are the robustness and predictability criteria for models used in impact studies (e.g., hydrologic impacts of land use and global change)?

The workshop contained energetic and substantial discussions. It was evident that the topic resonated with the interests of the research community and the demands of federal agencies and international research programs. There are currently a number of U.S. agencies—e.g., the interagency U.S. Global Change Research Program (USGCRP)—and international research programs—e.g., World Climate Research Program's (WCRP) GEWEX and CLIVAR—that identify predictability of hydrologic systems as being among their priorities. More specific agency examples include National Oceanic and Atmospheric Administration's GEWEX American Prediction Project (GAPP), National Aeronautics and Space Administration's Seasonal-to-Interannual Predictability Project (NSIPP), and National Weather Service's Advanced Hydrologic Prediction System (AHPS). Representatives from these and other agencies supporting COHS opened the workshop by defining their program requirements and objectives in predictability science related to the water cycle.

The report is divided into 4 chapters. Chapter 1 gives details of the workshop and provides the motivation for this present report. The definitions of three different types of limits-to-prediction are given in Chapter 2. Chapter 3 builds on the workshop presentations and frames the challenges in predictability science. Chapter 4 provides conclusions based on the workshop presentations and committee discussions. These conclusions present some of the promising scientific directions that could provide a starting point for either understanding the predictability of hydrologic systems or identifying what the limits to prediction are in these systems. The hope of the committee is that both the research and user communities find the discussions of this report useful both for implementing research strategies and for identifying the ways in which predictability research can be integrated more effectively into operational activities.

Dara Entekhabi, Chair
Committee on Hydrologic Science

Acknowledgment of Reviewers

This report has been reviewed by individuals chosen for their diverse perspectives and technical expertise, in accordance with procedures approved by the NRC's Report Review Committee. The purpose of this independent review is to provide candid and critical comments that will assist the institution in making its published report as sound as possible and to ensure that the report meets institutional standards for objectivity, evidence, and responsiveness to the study charge. The review comments and draft manuscript remain confidential to protect the integrity of the deliberative process. We wish to thank the following individuals for their review of this report:

J. D. Albertson, University of Virginia
David Ford, David Ford Consulting Engineers
Efi Foufoula-Georgiou, University of Minnesota
Wendy Graham, University of Florida
Charles T. Haan, Oklahoma State University
William Kustas, Agricultural Research Service
Roger A. Pielke, Jr., National Center for Atmospheric
    Research

Although the reviewers listed above have provided many constructive comments and suggestions, they were not asked to endorse the conclusions or recommendations nor did they see the final draft of the report before its release. The review of this report was overseen by Eugenia Kalnay, University of Maryland. Appointed by the National Research Council, she was responsible for making certain that an independent examination of this report was carried out in accordance with institutional procedures and that all review comments were carefully considered. Responsibility for the final content of this report rests entirely with the authoring committee and the institution.

Contents

1

Background and Goals

NRC COMMITTEE ON HYDROLOGIC SCIENCE 1999 REPORT

Predictions of hydrologic phenomena such as floods, seasonal precipitation deficit, aquifer response, subsurface contaminant dispersion, land use and global change impacts, etc. are practical ways of dealing with hazards. Predictions are often the foundations of hazards mitigation strategies. Predictions begin with a characterization of the current state of the system, i.e. initialization. Then the states of the system are predicted into the future according to our current understanding of the dynamic behaviors of the system. As a result observing systems and conceptual understanding are the engine behind operational prediction systems. In this way hydrologic science and operational hydrology work together to reduce the hazards faced by the public. Better characterizations of the system and more effective observing networks are needed to improve predictions. Predictability and limits-to-prediction are themes in hydrologic science that are at the core of both the research community and the operational field.

Predictability and limits-to-prediction in the hydrologic sciences emerged as a priority topic for COHS during the development of its first report on the hydrologic science content of the USGCRP plan. The NRC (1999) study titled *Hydrologic Science Priorities for the U.S. Global Change Research Program: An Initial Assessment* was published in September 1999 and it identified predictability and variability of regional and global water cycles as one of its two priority themes (together with coupling of hydrologic systems and ecosystems through chemical cycles). In NRC (1999) a series of specific science questions are posed in order to direct hydrology predictability research in a progressive direction. Addressing these science questions and bringing them to closure mark milestones in this research path.

NRC (1999) proposes that research in this area should be driven towards three main goals. The first goal is the identification of predictable patterns on all pertinent spatial and temporal scales in the water cycle. Because the water cycle is composed of components with varying memory (e.g., mean residence time in atmosphere is about nine days whereas in the active groundwater system the memory may be decades), there is a natural dampening and potential for predictability based on persistence. The prevalence of autoregressive and Markovian statistical models in hydrology are testament to the recognition that the dampening of signals in various component of the water cycle may be effectively harnessed to make useful predictions. It is now recognized that there may be even more opportunities for prediction if the climate system as a whole is considered. The oceans are the long memory components of the climate system and their influence on interannual to decadal climate variability is becoming more clearly understood. This understanding may be used to make long-range predictions of regional water cycle processes. The identification of predictable patterns and linking them to large memory processes is now emerging as a promising strategy for understanding predictability of hydrologic systems and is supporting the development of useful prediction tools. It is also recognized that memory in the regional climate systems may be due to factors other than the inertia of heat and moisture reservoirs. Establishment of positive feedback mechanisms may also have the same effects as memory by prolonging an event or excursion. Box 1.1 lists a number of specific science questions from NRC (1999) associated with this first goal of predictability research in the hydrologic sciences.

The second goal identified by NRC (1999) is understanding the sources of uncertainty and the propagation of uncertainty in hydrologic systems. Innovative monitoring and multi-source data fusion techniques are required to quantify variability and its

Box 1.1
Science questions on predictability and limits-to-prediction from the
1999 NRC report: *Hydrologic Science Priorities for the US Global*
Change Research Program:
An Initial Assessment

Distinguish the Predictable and the Unpredictable Patterns of Variability
• What type and location of measurements will most enhance predictability? To what extent is regional-scale hydrology predictable?
• Across which regions and seasons can predictability of regional water cycling be enhanced by robust coupled land-atmosphere modeling?
• What special physical and statistical features (e.g., process pathways, influences across scales) can be used to link large-scale climate and regional-scale hydrology in the case of extreme events and how are these features different for the case of floods and the case of persistent droughts?

Identify Sources of Variability and their Propagation in Hydrologic Systems
• What combination of remote and in situ observations and paleohydrologic records are required to identify shifts in regional and local hydrologic properties due to both natural and human-induced factors?
• Are there spatial patterns in the variability in the hydrologic record that may serve as reliable predictors of the impacts of global change?

Understanding the Scaling and Linkages of System Components
• At what scales and for which processes should the spatial structure of surface heterogeneity be incorporated into the upscaling strategy for hydrologic models?
• What physical constraints arising due to coupling of water and energy cycles with other systems may be used to bound the estimates of local and regional hydrologic cycles?

changes with scale. The focus of the specific science questions associated with this goal (see Box 1.1) is on the quantitative characterization of the impacts of perturbations in hydrologic systems. These perturbations may be due to human-induced factors such as land use change or they may be due to natural variability in climate forcing.

The third and final major goal posed in NRC (1999) for predictability research in the unique context of hydrologic sciences is to understand how variability in hydrologic processes change with spatial scale. Heterogeneity in landscape properties (e.g. geology, ecology, terrain) are ubiquitous in hydrology. How processes such as water, energy, and biogeochemical fluxes are dependent on the variabilities in these properties needs to be addressed to advance understanding. Box 1.1 also lists a number of specific science questions for this goal.

The task of developing the full science and implementation plan for the USGCRP element went to the sixteen-member Water Cycle Study Group, organized by the federal agencies and chaired by Dr. George Hornberger (University of Virginia).

THE USGCRP WATER CYCLE INITIATIVE

In June 2001 the USGCRP Water Cycle Study Group published a comprehensive report titled: *A Plan for a New Science Initiative on the Global Water Cycle* (Hornberger et al. 2001). The report poses several priority science questions as well as several specific goals for predictability research that are listed in Box 1.2.

In the USGCRP report, a series of research needs are articulated that can be mapped to activities that will have to be undertaken by the appropriate federal agencies. A sampling of these needs indicates the wide scope of the efforts.

A new program is needed in the science and mathematics of water cycle predictability to guide applications of atmospheric and hydrologic theories over a broad range of space and time scales. Climate predictions on seasonal and longer time scales must be made within a probabilistic framework that takes into account the uncertainty of initial and boundary conditions, as well as the inherent characteristics of the distribution of possible states that

Box 1.2
Key elements of *A Plan for a new Science Initiative on the Global Water Cycle*
(Hornberger et al., 2001)

Priority Science Questions

1. What are the underlying causes of variation in the water cycle on both global and regional scales, and to what extent is this variation induced by human activity?
2. To what extent are variations in the global and regional water cycle predictable?
3. How will variability and changes in the cycling of water through terrestrial and freshwater ecosystems be linked to variability and changes in cycling of carbon, nitrogen and other nutrients at regional and global scales?

Specific Goals in Predictability and Limits-to-Prediction in Hydrologic Systems

1. *Demonstrate the degree of predictability of variations in the water cycle on a range of time scales (daily to centennial).* This goal is to be reached through a number of program elements that include: nested modeling to deal with scales interactions, probabilistic modeling to deal with uncertainty in modeling, and scaling models for interpreting observations.
2. *Improve predictions of water resources by quantifying fluxes between key hydrologic reservoirs using observations, process understanding, and numerical modeling.* The program elements required to reach this second goal include observations using surface networks: precipitation, basin-scale recharge that links the surface and subsurface reservoirs, stream-aquifer interaction which also is related to these two reservoirs, and evaporation that links the surface and the atmospheric reservoirs.
3. *Establish a systems modeling framework for making predictions and estimates of uncertainty that are useful for water-resources management, natural hazards mitigation, and policy guidance.* The program elements for this goal include transfer of information from physical to socioeconomic models.

may ensue from the given initial state. Research is required to place current ad hoc methods of producing ensemble model predictions on a firmer theoretical basis.

A systematic approach to model design and development is needed that will permit determining the scales at which predictive information should be exchanged within a nested modeling approach. This research will be heavily computational, requiring enhancements to available national computing capabilities.

The development of coupled land-atmosphere models should be accelerated through the better use of data assimilation techniques. One existing vehicle for this development is the new U.S. multiagency initiative known as Land Data Assimilation System, or LDAS. LDAS should be supported and expanded to include data representing snowpack and high-latitude glaciers. Studies should examine whether two-way land-atmosphere coupling or climate modulation by local hydrologic processes results in predictability that can be exploited through coupled modeling. For seasonal and longer lead prediction of water fluxes, a modeling strategy must be developed to minimize the propagation of uncertainty among the components of such predictive models.

Field campaigns and intensive observation programs to better understand interactions among land, ocean, and atmosphere are needed to isolate the effects of fast and slow processes in the hydrological cycle. Enhanced field campaigns should take place over multiple years to observe large-scale surface conditions, surface fluxes, and atmospheric variables. These large-scale observations would be supplemented with simultaneous observations of the slower components of the land system, such as groundwater levels.

A continuing effort to use observations to close water budgets is critical. New data sets geared specifically for budget studies are needed. Because analysis budgets are the main link between models and observations, they should be rigorously tested against all observations, especially those hydrometeorological observations developed to cover broad space and time scales. New continental and global hydrometeorological data sets will be required to support these activities. These data sets include gridded (or equivalent) observations of streamflow over continental domains, and gridded high-resolution precipitation data. Expanded budget studies covering snow accumulation, melt, runoff, and evaporation of snow in continental regions should also be undertaken to understand how snow contributes to the water cycle.

Satellite observations are needed for hydrological variables not yet remotely sensed, and for which technology development may be required. These include surface soil moisture at high resolution (~10 km for hydrometeorology applications, ~40 km for hydroclimatology applications), surface freeze/thaw condition, diurnal cycle of precipitation, river and water bodies altimetry, and snow cover and water equivalent.

An initiative should be designed to integrate users needs into the development of the research agenda and to ensure that research results are provided in a form useful for users. Estimates should be developed of the natural variability of surface hydrological processes that can be incorporated into water resource systems design and management, with reduced dependence on historical observations. Ensemble forecast products for operating water resource systems should be produced, with a primary focus on reservoir systems (or, in some cases, free-flowing rivers), but with implications for groundwater in systems that conjunctively use surface water and groundwater. Model testing facilities should be established at existing weather and climate prediction centers (like National Centers for Environmental Prediction (NCEP)), which would be charged with facilitating model evaluation and the transfer of methods from the general research to the operational modeling community and vice versa.

THE COHS WORKSHOP

These two reports (NRC 1999 and Hornberger et al. 2001) summarized above provide context for material presented at the COHS workshop in Boulder, Colorado. The workshop additionally served to engage the larger community outside of these two study groups. The primary goal was to start along the path defined by the reports by making contributions in three main areas: 1) provide concise definitions for predictability and limits to predictability, 2) identify the key technical challenges for research advances in predictability science, and 3) make recommendations to federal agencies in implementing a research program in this topic. These three contributions are each contained in the sections of this report.

In preparing for the workshop it became necessary to choose a particular example application from the diversity of topics in hydrologic science to demonstrate the current status and future prospects for predictability science in hydrologic systems. Predictability and understanding the limits-to-prediction are core elements of research in many specializations in hydrology. Subsurface flow and transport, surface water hydrology and chemistry, ecohydrol-ogy, snow hydrology, hydrometeorology and hydroclimatology as well as other topics in hydrologic science all involve predictions and predictability in one form or another. COHS decided that it is preferable to choose one context and delve deeply into it in order to identify some of the structural issues associated with predictability research. Such structural issues include the criteria for posing science questions and for defining metrics of progress. In the workshop the context of hydrometeorology a hydroclimatology was used to introduce some of the issues associated with predictability research. Nonetheless the findings are applicable to most other contexts in hydrologic science.

In preparing for the workshop it also became necessary to recognize that the evolution of predictability science and operational prediction systems are intimately linked. During the workshop it was recognized that the two are complementary and that together they form a synergistic approach to advancing understanding in hydrologic science that is valuable to applications. Predictability science and the quest to define the limits-of-prediction may be directed towards the goal of gaining basic understanding of hydrologic systems. Alternatively predictability science may be directed towards the goal of improving operational predictions in the context of an application. There are circumstances possible where the two goals are aligned together and they cannot be separated. Nonetheless the two goals for predictability research are distinguished here because they often require different driving science questions, data, and metrics of progress.

Progress in predictability research directed towards improving operational prediction is measured in terms of increased prediction accuracy or forecast skill. Such improvements may be achieved by empirical means that are useful in the application context but do not necessarily provide added insight for guiding the research into the future generation of prediction systems. Similarly predictability research directed towards basic understanding may strive to understand a fundamental feature or define a new paradigm that promises to become the basis for an improved operational prediction system at a later stage. In this context the goal is less to improve the immediate forecast skill but to develop the foundations for a future system. In fact, in the short term there may be a drop in forecast skill as a new paradigm is introduced but the paradigm shift sets a new trajectory that promises great gains (see Figure 1.1).

SUCCESS AND FAILURE IN APPLYING ADVANCES IN PREDICTABILITY SCIENCE

Because policy decisions are often forward-looking, they are, in part, based on predictions. In the Workshop Roger Pielke, Jr. discussed the makings of an effective research and application program. Under this model there is a parallel undertaking of research and use of predictive models, linked by the communication between researchers and users of predictions. Success in prediction depends on the effective communication between these groups, leading to predictions which, 1) provide the most skill possible given the available data and modeling capabilities, 2) account for the stated needs of prediction users, and 3) have effectively communicated and understood uncertainties.

Because the linkages among research, prediction and use involves a series of interacting processes, characterizing success or failure is difficult. Absolute success in the cycle occurs when a skillful prediction is communicated efficiently, and used effectively in formulating a decision that has a value to society. Users communicate their needs to predictors, and the predictors communicate the predictions and the uncertainties associated with the prediction to the users. The users are then able to make decisions, which are beneficial to society. The 1997-1998 El Nino event in California, illustrates the effective interaction between the users and predictors. During this period the users and the predictors communicated the importance of identifying the timing and

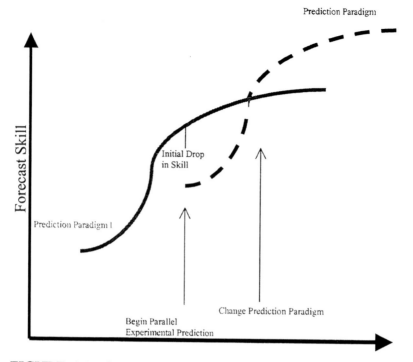

FIGURE 1.1: Conceptual figure demonstrating how advances in predictability science transition to improved operational predictions. The effectiveness of the predictions is measured with a skill score. Prediction systems are based on existing paradigms or scientific understanding. Initially the system has a slow rate of increase in skill. Errors in implementation, uneven completion of auxiliary systems, and gradual training of personnel in the prediction system some of the reasons why the initial increase in prediction skill can be modest or even negative. As the prediction system matures it undergoes a period of rapid improvement in its effectiveness. As the prediction system and its supporting science paradigm mature, the system again experiences slower rates of skill increase with time. In this phase the prediction system has essentially reached it highest potential for characterizing and predicting hydrologic phenomena.

potential impacts of the anticipated El Nino. Through research programs, assembled data and established models, researchers were searchers were able to predict the upcoming El Nino and were also able to determine that it would cause heavier than normal precipitation in California. This prediction was ably communicated to decision makers, who were involved in the process throughout the prediction. Because of the interaction between researchers and users, effective decisions were made and communicated at a local, regional and national scale. As a result, communities were able to prepare for and, in some cases, mitigate the potential for disaster. Although the flooding did cause significant damage, the interaction and communication between the policy makers and the researchers is believed to have limited the damage from this event.

It is often more difficult to characterize failure in the system. Effective decisions may be made despite poor prediction and skillful predictions do not always result in decisions with a value to society. This breakdown is often a result of poor communication between predictors and policy makers.

Catastrophic flood events exact a high toll in lives and property every year. For example the 1997 Red River Flood provides can be used to illustrate the breakdown in the cycle, where failed communication between predictors and policy makers resulted in a disastrous outcome. During this event, predictions were made for record-river flooding to take place. Forecasters gave a deterministic forecast with a value of 49 feet for the flood crest, which was slightly greater than the previous record crest for the river. This prediction did not effectively convey the associated uncertainties, rather indicated a fixed value for the flood crest, which was only marginally larger than the previous high water mark. The river crested at a height of 54 feet, resulting in extensive damage to the unprepared communities, which had prepared for river crests of no greater than 49 feet. (It should be noted that predicting a new record crest level is a remarkable achievement in operational prediction.)

Although many elements of the hydrologic system offer varying degrees of predictability, it is crucial, as evidenced from the above examples of success and failure, that the uncertainties and the limits-to-prediction be established and well communicated to user groups.

2

Predictability Science: Definitions

During the workshop, three types of predictability and limits to prediction were identified. The first two follow from classical definitions and the third one introduced in this workshop is a particularly relevant addition for hydrologic systems.

Predictability of the first kind is associated with information present in the initial conditions. Its importance can be assessed by the extent to which the predictions are sensitive to the initial conditions. For example quantitative precipitation forecasts (QPF) remain major challenges for the community. Various approaches to QPF are each characterized by varying performances at different lead times and aggregation levels. A complete QPF system will re-quire not only a systematic framework for the merging forecasts from various approaches but also deeper insight into precipitation, severe weather, and runoff processes that ultimately produce floods, which are the most costly (in terms of human life and property) realization of weather hazard in many places including the U.S. Sensitivity to initial conditions is also charac-teristic of many other hydrologic systems (e.g., surface-subsurface processes, integrated eco-logical-hydrologic systems) that involve coupled processes where two-way linkages result in feedbacks that can amplify errors in initialization.

Predictability of the second kind is associated with information present in the boundary conditions. Its importance can be assessed by the extent to which predictions are sensitive to these boundary conditions. For example there are known local and remote factors (such as ocean surface temperatures) that affect variability in regional precipitation on intraseasonal to interannual time scales. Conditioning long-lead predictions on slower evolving states of the climate system has been shown to only partially reduce the uncertainty of forecasts. There is growing demand for long-lead predictions that reduce the risk associated with climate-sensitive activities. Robust operational prediction systems that may meet this demand are built on the basic understanding of how local and remote factors contribute to the total hydrologic variability. Increasingly ensembles forecasts are used to identify the bound on the uncertainty associated with error-prone parameters and inputs. Examples of needed basic research in this area include 1) increasing the reproducibility of ensemble models, 2) developing statistically representative ensemble members based on randomized inputs, and 3) developing ensembles of cases from each in an ensemble of models.

Hydrologic systems contain heterogeneous geological, topographic, and ecological fea-tures that vary on multiple scales. The pervasive nature of nonlinear scale interactions in hydrological systems was introduced in the Workshop as a third source of predictability (or, in most cases, loss of predictability). Predictability of the third kind asserts that the effective response of systems at larger scales is not completely determined by scaling local processes (e.g., scaling up from small scales to larger scales is not a linear process). In hydrologic science heterogeneity is a rule and it cannot necessarily be fully captured by randomization of parameters. Interactions among microscale features often lead to effects that are not completely represented in macroscale predictions based on effective parameters for microscale models. Examples include enhanced surface flux due to land-breeze circulations over heterogeneous patches, on regional recharge and discharge patterns over complex terrain. There are processes and conditions when the effective parameter approach to scaling may be feasible. In the remaining circumstances the macroscale and microscale predictive relations for hydrologic processes may have different functional forms and dependencies. Furthermore, there may be organizing principles at work that result in simple procedures for statistically relating variables across a wide range of scales in the hydrologic system. These scale considerations affect limits to prediction in hydrologic systems and they place in question traditional ideas in hydrologic predictions. For example, spatial and temporal averages are not necessarily more predictable as traditionally believed if the averaging covers a scale that contains a strong transition or change in behavior (analogous to a bifurcation in dynamic systems).

3

Challenges in Predictability Science and Limits-to-Prediction for Hydrologic Systems

In this report, we derive challenges for predictability science examining knowledge gaps and operational needs. For each challenge we identify milestones to mark progress. Finally, we limit discussion to the subfield of hydrometeorology out of a practical necessity, with an implied relevance to the broader discipline of hydrology.

The coupled hydrometeorological predictability problem is chosen to show how a focused set of challenges in predictability science and "limits-to-prediction" research may be derived from current gaps in understanding and from the current operational needs. More importantly, milestones need to be defined for these challenges in order to guide progress in purposeful directions and establish two-way communications between the research and applications communities. Predictability and understanding the limits-to-prediction are, however, core issues in many specializations in hydrology such as ecohydrology, subsurface chemical fate and transport, etc. Advances in understanding predictability and using that knowledge to improve predictions benefits the nation in a wide range of applications (Box 3-1). The purpose of this section, however, is to focus on one of the predictability problems in hydrologic science—hydrometeorology—and in the process identify those structural issues that are common to diverse hydrologic science applications.

During the past century, many anomalous climate events

Box 3-1
From Research to Operations

Considerable operational and regulatory hydrology is based on predictive model output. Hydrologic models are essential tools used to characterize the benefits and costs of proposed private and public actions.

Simulation models used in hydrology attempt to predict the movement of water, chemicals, and sediment across the landscape. In this respect there are a number of problems related to the setting—i.e., the natural and built landscape—that are common to all hydrologic models. These include underresolved processes (subgrid scale effects associated with discretization of processes that vary on a wide range of scales), incomplete chemical and biological parameterizations (shortcomings in characterizing and monitoring heterogeneous chemical, surface chemistry, and microbial processes in the environment), and lack of adequate sampling for model specification and initialization (failure of sparse monitoring networks to capture the true variations across the landscape). Each type of hydrologic model, nonetheless, has additional problems that are specific to its construct and application context.

It is not feasible to perform a complete survey of research needs for all predictive hydrologic models and prediction applications in the various agencies engaged in these activities. Here we present three examples of application areas outside of hydrometeorology where research in predictability and limits-to-prediction may have practical impacts and may serve the nation by enhancing the capability of predicting environmental processes linked to the movement of water across the landscape.

Erosion and Sediment Transport (USDA)
Problem: We cannot currently predict, to even within orders of magnitude, how many tons of sediment leave a watershed in a year.

Critical Issues: The critical issues include development of mechanistic models of erosion to replace empirical predictive models such as those based on the widely used "Universal Soil Loss Equation" and development of models for sediment storage on the floodplain and resuspension and transport processes in the river.

Watershed Rainfall–Runoff Transformation (NWS, USGS)
Problem: We cannot currently predict the spatial pattern of watershed response to precipitation and cannot quantitatively describe the surface and subsurface contributions to streamflow with enough accuracy and consistency to be operationally useful.

Critical Issues: Initial and boundary conditions are the critical issues. Watershed runoff and streamflow are affected by heterogeneity in soil hydraulic properties, landscape structural properties (e.g., hydrogeological layering, compaction of soil horizons, and soil organic content, roots, and pores), soil moisture profile, surface–subsurface interaction, interception by plants, snowpack, and storm properties. Although our understanding of individual processes is improving, the integration of that body of knowledge in spatially distributed predictive models has not been approached systematically.

Groundwater Management in Irrigated Agriculture (USDA, USGS, BLM)
Problem: We need to provide sufficient water for the crop plants and to minimize movement of harmful chemicals to the aquifer below.

Critical Issues: We do not have reliable means to go from core or plot scale measurements of hydraulic properties to estimated field scale hydraulic pathways in predictive models. We need to replace empirical parameterizations of chemical and biological fate and transport in the environment with models based on the results of fundamental understanding in heterogeneous chemistry, surface chemistry, and microbiology.

have disrupted American lives. Persistent droughts like those associated with the dust bowls of the 1930s and the recent drought of 1988 ruined Midwest crops and farmland. The Mississippi floods of 1927 and 1993 were equally devastating. Even larger regional climate variations may occur in the future, especially if the global climate is seriously influenced by the rise in concentrations of greenhouse gases, as some models and observations indicate. Increasing the extent to which such events can be predicted is an integral component of the World Climate Research Programme's (WCRP) Global Energy and Water Cycle Experiment (GEWEX) as well as the U.S. Gloabal Change Research Program (USGCRP) Water Cycle Plan, National Aeronautics and Space Administration's Global Water and Energy Cycle (GWEC), and Natiohnal Oceanic and Atmospheric Administration's Office of Global Program GEWEX Americas Prediction Project (GAPP) programs.

CONCEPTUAL MODEL FOR MAKING PREDICTIONS

The conceptual framework for hydrometeorological predictability is that of a coupled land–atmosphere–ocean system. At the largest time and space scales, the cycling of water over continental regions can be viewed as net inflow of water vapor to a particular basin, net transfer of water from the atmosphere to the surface by excess of precipitation over local evapotranspiration, and net river discharge from the basin, which over the long run must balance the net inflow of water vapor. On a continental scale, the rivers discharge into the oceans, and the oceans are net sources of atmospheric water, which is transported to the land regions. The fluxes and stored amounts of water vary greatly in space and time. Parts of these variations are regular, following the annual cycle of solar forcing in time and the physical controls of geography (topography, soil, and vegetation cover) in space. Superimposed upon these regular variations are the irregular fluctuations or changes caused by the chaotic dynamics of the land–atmosphere–ocean system. Such chaotic behavior is generated both internally in the basin and externally (e.g., by the general circulation of the atmosphere). Storage processes within soil and vegetation modulate both the regular and the irregular variations in water and energy fluxes.

During the workshop, Huug van den Dool reported on empirical studies that indicate the tendency toward persistence of summer droughts and that suggest early-season soil moisture anomalies as a contributing causal agent. Recent extreme hydroclimate events have provided a focal point for studies of land–atmosphere interactions, illustrating the complexity of the atmospheric response to surface anomalies. The heavy precipitation that caused record-breaking flooding within the Mississippi River basin in 1993 has been associated alternatively with high and low soil water anomalies in different areas. Physical processes invoked in the alternative explanations include (1) surface-heating effects on the boundary-layer capping inversion and associated suppression of deep convection, (2) influences of surface conditions on the low-level jet in the southern Great Plains. At the other hydroclimatic extreme, initially dry soil conditions in the Mississippi basin have been put forward as a possible cause of the 1988 summer drought.

It appears that potential atmospheric predictability associated with land-surface anomalies could be especially significant during the warmer part of the year. During the workshop, Randy Koster summarized studies (using climate models) that indicate that land factors are contributors to seasonal precipitation variability under a set of conditions that favor strong land–atmosphere coupling. Outside of these circumstances, the variations in seasonal precipitation do not appear to be related to antecedent or concurrent conditions at the land surface. Remote influences, such as seasonal to interannual ocean temperature anomalies (e.g., El Niño-Southern Oscillation), probably outweigh any land influences during winter and could also be important in setting up the initial springtime land surface soil moisture anomalies. Similarly, Hong and Kalnay (2000) showed that the drought of 1998 over Oklahoma and Texas, once established in early spring by sea surface temperature (SST) anomalies and by favorable initial conditions, was maintained by local soil moisture positive feedback. However, the influence of all hydrologic anomalies (soil moisture, snow extent, soil freezing and thawing) on subsequent weather patterns and short-term climate predictions is still largely unknown. Observing, understanding, and modeling these coupled hydrometeorological processes through the full range of spatial and temporal scales are essential not only for developing long-range predictive capability, but also for developing basic understanding of the water cycle. During the workshop, Kevin Trenberth emphasized the synergy between models and observations and the water and energy cycles in addressing one of the key issues in global change research today: Is the hydrologic cycle changing?

Three important unresolved issues are (1) whether the coupling between the land surface and the climate system is sufficiently strong so that knowledge of the land surface states will enhance prediction, (2) whether the accuracy and resolution of current and future remote sensing observations are sufficient to provide information useful for enhanced predictions, (3) and whether current coupled models are capable of simulating critical processes. In other words, if we can accurately observe initial land surface conditions, does this result in increased hydro-meteorological predictability? When and where is this predictability likely to be most important? Are present global and regional climate models capable of simulating and hence predicting coupled features, and where are the current limitations? The talks by Adam Schlosser and Dag Lohmann at the workshop illustrated the range of activities in the hydrometeorological research community that focus on these questions with a variety of approaches and atmospheric and land surface models.

KEY UNRESOLVED ISSUES AND
RESEARCH CHALLENGES

Using predictability in the hydrometeorological system example as the context to introduce some key unresolved issues, five challenges are identified in predictability science and limits-to-prediction for hydrologic systems. Each is discussed below along with associated research milestones identified at the workshop.

Separating the Predictable and the Unpredictable

Although weather is not predictable beyond a few weeks because of its inherently nonlinear and chaotic nature, aspects of climate may be predictable for much longer because of the presence of low-frequency interannual variations such as El Niño, the Madden Julian Oscillation, and the Quasi-Biennial Oscillation. The term "potential predictability" is often used to define that part of the climate variations that exceeds weather noise. The idea is that the variance of seasonal precipitation is made up of a component reflecting daily weather (high-frequency random) variations, which are unpredictable beyond the deterministic predictability limits of about 2 weeks. The second component is any additional or multiplicative variance that is, at least, potentially predictable because of it links to physical systems with longer-range memory (e.g., oceans, continental soil moisture). The first component is considered noise and is estimated from a statistical model that is fitted from daily, within-season precipitation. Estimates of the climate noise are compared with the total variance, and where the total variance exceeds the estimated noise, one can conclude that there is a potential for long-range prediction.

The separation of weather noise and climate signal should not be interpreted as simply the separation of the effects of initial and boundary conditions for the land–ocean–atmosphere system. During the workshop, Roger Pielke, Sr., presented a case example where the coupled hydrology/ecosystem and climate model evolves into different equilibria depending on the initial conditions. Thus, the slowly varying component of the system (the ecosystem in this case) affects the system through both boundary and initial state effects.

Similar signal and noise separation techniques are required for other applications in hydrology where predictability associated with persistence, local factors, and remote influences needs to be separated.

The research milestones related to separating the predictable and the unpredictable include estimation of the time scales over which hydrologic variables can be predicted in the real world and determination of how well these time scales are simulated in modeling systems.

Characterization of Subgrid Scale (SGS) Processes

The massive development of computational resources has led to important advances in prediction methods and to a basic understanding of complex environmental phenomena. Nevertheless, the scale disparity intrinsic to linked hydrologic systems (e.g., land–atmosphere, surface–subsurface, water–ecosystems–biogeochemical, etc.) makes direct numerical simulation impossible. The systems contain variabilities on ranges of scales that are impossible to resolve on a common computational grid. As a result there remain fundamental issues in the formulation of computational models, especially in terms of the representation of subgrid scale (SGS) processes.

In the workshop, Joe Tribbia introduced an example that demonstrated the effects of SGS parameterizations on error propagation in models and demonstrated how model-based estimates of limits-to-prediction are affected by the approach to representing SGS processes. The example of the sensitivity of atmospheric forecasts to the SGS representation of moist convection and the formation of precipitating clouds showed that model results are highly dependent on subtle and buried assumptions. Looking at model results using two different SGS parameterizations provides a useful diagnostic tool for identifying discrepancies owing to SGS processes. Very high-quality observations are needed to choose the better of the two methods or to improve them, as Dr. Tribbia showed in the case of a forecasted Gulf Coast storm event.

SGS processes remain the Achilles heel of many prediction systems. They represent a limit to predictability (as defined in this report) brought about by the influence of processes with disparate scales. There are a number of other examples where SGS processes affect the behavior and skill of the parent model. These include the representation of atmospheric boundary layer and of its effects on the exchanges of moisture, heat, and biogeochemical substances between the surface and the atmosphere. The representation of pore scale to plot scale variability in soil characteristics has a large impact on the prediction of water movement near the surface. At even smaller scales, the molecular diffusion scale to pore scale characterization of soils significantly affects the prediction of large-scale contaminant plume migration in the subsurface. As demonstrated by Dr. Tribbia at the workshop, it is the availability of high-resolution and reliable observation data sets resulting from operational networks or field experiments that can identify discrepancies in predictions due to SGS processes.

Research milestones related to the characterization of SGS processes include the following:

• development of systematic ways of defining the relevant scales for a problem and how to pose models of the right complexity

• designing experiments that explicitly help formulate models for unresolved scales

• use of a hierarchy of models with differing levels of complexity, and use of ensembles of model simulations that include uncertainty to explore predictability

• systematic investigation, quantification, and cross-comparison of model sensitivities in well-posed and well-directed intercomparison projects

• estimation of the degree to which the atmosphere, the land-surface processes, and the subsurface are coupled together in the real world, including determination of how the degree of coupling differs among modeling systems.

Benefiting from the Synergy of Models and Measurements

Data assimilation or the merging of models and data is the application of the set of mathematical techniques that provides physically consistent estimates of spatially distributed environmental variables. Inverse problems are closely related to data assimilation and share many features. Estimates from data assimilation are often based on merging scattered and/or indirect measurements of states and parameters with dynamic models that impose physical consistency constraints. In this respect data assimilation is an effective strategy for extracting value (or information) from measurements that may be incomplete or noisy by themselves but form effective constraints on models that can provide the connectivity in space and time in between measurements. Because of its joint use of observations and models, data assimilation can also provide efficient tools to capture the multiscale variations of spatial fields in hydrologic systems.

A key consideration in data assimilation is that the models that provide the so-called background predictions and the systems that provide the same measurements are both uncertain. The role of data assimilation is to merge these two estimates based on their degree of uncertainty and produce a combined estimate that has desirable statistical properties (e.g., unbiased, minimum variance, etc.). Model calibration, which has a rich history in hydrology, is distinct from data assimilation in that the former is focused on the model and the latter on measurements and inference of the system state. More importantly, data assimilation is directed toward finding the errors of estimation given all the sources of uncertainty, whereas model calibration does not typically consider these uncertainties directly. The development of a data assimilation framework for hydrologic systems remains a major challenge (1) because of strong nonlinearities in the dynamic behavior of hydrologic systems, (2) because of involvement of diverse spatial scales in the determination of hydrologic events, and (3) because of a lack of reliable knowledge about the uncertainties of measurements and models.

During the workshop, Baxter Vieux presented case examples of predictability research (basin runoff prediction) that show the need for developing the synergy provided by advanced models and intensive measurements. In order to benefit from the synergy of models and measurements both in the context of data assimilation and in general, it is necessary that models be physically based and well tested. Classic hydrologic models that have been optimized for use with point observations (such as precipitation and streamflow) are inadequate for extension to data assimilation which is distributed in space. What is needed are models and model components for hydrologic processes need to be developed that can work well with point as well as with mapped observations.

Models must also follow predefined criteria for parameter parsimony and overall observability. Observability is a property of a system that is defined as the degree to which an increasing number of observations leads to diminishing uncertainty about the state of the system. Finally, it should be recognized that models for prediction and for assimilation applications may have significantly different requirements/characteristics. The requirements for models used in prediction and these for models used in assimilation need to be defined.

Research milestones related to synergy between models and measurements include the following:

- improve the initial conditions (and thus take better advantage of predictability associated with information present in the initial conditions);
- allow efficient improvement of the models by comparison of short forecasts with the observations
- provide community data sets on regional hydrologic systems
- extend data assimilation systems to take advantage of emerging satellite data.

Making the Observations that
Accelerate Model Improvements

Accurate, appropriate ground-based measurements of both the state of hydrologic reservoirs and fluxes between reservoirs are the single most critical factor that will drive advances in predictability and predictions. These critical areas illustrate compelling needs. First, distributed, well-designed networks that measure temperature, precipitation (rainfall and snowfall), snowpack, soil moisture, vegetation properties, radiation, wind, evaporative flux, and humidity will provide the foundation for improved predictions of water fluxes at or near the land surface. Precipitation is one of the key forcing factors of regional hydrologic systems. During the workshop, Witek Krajewski provided an overview of the current state of precipitation- monitoring systems and the prospects for the future enhancement of the networks. Though some capability is currently available in the United States and worldwide, to address each of these measurements, significant improvements will be required both nationally and internationally if we are to achieve advances in predictability of the water cycle at the land surface. Similarly, new measurement technology and network design are critically needed to improve the predictability and prediction of chemical fluxes, of transportation, and of impacts on terrestrial ecosystems and aquatic ecosystems in both inland and coastal waters. Finally, new measurements to characterize properties of the earth's "critical zone" are sorely needed for both hydrologic science and integrative studies linking hydrology with other earth and environmental sciences. The critical zone is "the heterogeneous, near-surface environment in which complex interactions involving rock, soil, water, air and living organisms regulate the natural habitat and determine the availability of life-sustaining resources" (NRC, 2001a). Measurement networks must be well designed for emerging research and applications; they cannot simply be extensions of existing networks. Designs that served the predictive tools of past decades may not be the most appropriate for integrating ground-based and remotely sensed measurements for the predictive tools we will have available in the coming decades.

Better process understanding is the key benefit of intensive field campaigns and sustained research at long-term experimental sites. Although the hydrologic community can point to a number of successful field campaigns lasting from days to months, the availability of long-term experimental sites has been limited. Both the U.S. Geological Survey and the U.S. Agricultural Research Service have well-established small-scale research catchments and have a long-term commitment to maintaining them as research sites. However, these only address hydrologic and biogeochemical issues at scales of a few meters to a few kilometers. The National Science Foundation's Long-Term Ecological Research (LTER) program is also a resource for understanding processes in the critical zone at scales of a few kilometers to tens of kilometers. Efforts such as NASA's and NOAA's Continental-Scale International Experiment (GCIP) and GEWEX America Prediction Project (GAPP) efforts have addressed important hydrologic issues at the regional to continental scale. However, there is currently a critical need for sustained investigations at these larger scales as well as at smaller scales.

During the workshop, Vijay Gupta stressed that sustained, long-term regionally representative ground-based measurements in research basins can serve as a test bed for the testing of new scientific hypotheses and instruments. As hydrologic science develops further applications of remote sensing, requirements for ground-based network design are changing. For example, the current network of index sites for snow water equivalent may not be ideally located to provide ground data combined with snow-cover area from satellite remote sensing, to estimate basinwide snow water equivalence. Rather, a network of sites with greater topographic variability in siting may be more appropriate. Simulation and design studies with dense measurements from research basins can be used to evaluate tradeoffs and demonstrate data value. Research milestones related to making observations that accelerate modeling progress include the following:

• development of benchmarks for monitoring systems, and implementation of special initiatives in algorithm development and assessment of new technologies
• estimates of uncertainty associated with observations made at different scales and using different measurement technologies
• improvement of access to existing data
• development of a multiagency definition of hydrologic data requirements, development of strategies for coordinated observations, and development of effective mechanisms for data sharing and dissemination
• estimation of evaporation and recharge on scales that allow linking the subsurface, surface, and atmospheric hydrologic systems
• use and promotion of paleo/proxy data for insights on long-term variations.

Measuring Predictability

Designing metrics for quantifying limits-to-prediction is a major challenge. During the workshop, Upmanu Lall provided a lecture on the state of the art in quantifying predictability and pointed out some of the more promising future pathways. He showed that past efforts aimed at quantitatively determining predictability in hydrologic systems have been based either on idealized systems of dynamic equations or on mechanistically and/or numerically refined model studies believed to be representative of the "true" system. In both cases, the approach has been to quantify predictability intrinsic to the system by employing analogs that enable the use of methods such as those of nonlinear dynamics (e.g., Lyapunov exponents and information theoretic/entropy) and/or statistical comparison of model forecasts with observations (e.g., threat scores and root mean squared errors). However, we are still unable to adequately deal with many predictability measure issues critical to hydrology.

In the workshop, Efi Foufoula-Georgiou used the example of Quantitative Precipitation Forecast (QPF) to demonstrate the difficulties in defining robust and reliable measures for limits-to-prediction. These difficulties mostly have to do with the range of scales over which hydrologic processes vary and with the intermittency in some of the variables. Traditional metrics such as root mean squared error or threat scores to often fail to adequately capture the accuracy of predictions.

Metrics that increase the understanding of the various kinds of predictability and that help to characterize memories, pathways, and feedback in hydrologic systems are needed. Further, persistence effects need to be distinguished from other factors that lead to predictability (e.g., remote influences and feedback mechanisms). One approach that can address the above issues is to define a probabilistic framework that includes the concept of a combined stochastic-deterministic error and allows for its quantification. Such a framework would permit the development of numerical optimization strategies such as adaptive mesh refinement and data assimilation that are consistent with the level of accuracy justified by the available data.

During the workshop, Roger Ghanem presented one such framework. The framework maintains that a predicted variable has multiple sources of error. There are computation-related errors that can be controlled by refining the numerical approximations. There are parameter uncertainty errors that can be controlled through refining a probabilistic approach to model parameters. There are also initialization and boundary specification errors that can be controlled through refining the measurements used in prediction models. Finally, there are model structure errors that can be refined through improvements in understanding the modeled processes. The challenge is to develop techniques for separating these errors and controlling them individually.

As yet, the hydrologic science community has no commonly agreed-upon or widely used techniques for evaluating forecast skill that are robust with respect to other factors such as persistence and intermittency. For hydrologists, a great remaining research challenge is to properly design tools and techniques so that the predictability and limits of prediction in hydrologic systems can be better quantified. These measures need to be defined in the context of specific forecast quantities and spatial and temporal scales.

Research milestones related to measuring predictability include the following:

• definition of robust measures of limits-to-prediction that account for scale and inter-mittency issues and that are capable of distinguishing persistence effects

• introduction of methods to infer predictability and limits-to-prediction from observa-tional data sets.

4

Conclusions

The workshop presentations and discussions confirmed the importance of developing an understanding to the limits of hydrologic prediction. Discussions during the workshop and written contributions by the participants resulted in defining milestones of progress in advancing predictability research and understanding limits-to-prediction in the hydrologic sciences. These milestones are valuable for any research initiative because they define the important and critical research directions. Additionally, they will allow the development of a timeframe for progress by the research program. In addition, such objectives are benchmarks for tracking the maturation of a research area, the vision for advancement of the area, and the metrics for progress.

While it is recognized that USGCRP agencies have focused research activities on forecasting and prediction, the workshop participants indicated that USGCRP agencies should establish programs to investigate the limits to predictability of the wider range of hydrologic variables. For example, current programs tend to focus on meteorological prediction, while understanding the limits-of-prediction for groundwater contaminant transport or ecosystem dynamics have received less attention. Yet these systems are of critical importance to the nation. In fact, the NRC report on environmental grand challenges (NRC, 2001b) included improved hydrologic forecasting among five priorities in environmental science.

The workshop identified the need for furthering the understanding between predictability and sub-grid-scale processes. Recent research suggests that increased resolution of distributed hydrological models has not necessarily lead to improved predictions due to the fact that the lack of understanding and modeling of sub-grid scale processes is not compensated by improved resolution data sets. The increased availability of high-resolution data sets (e.g., data from space-borne sensors) allows for research programs that address the relationship among distributed data sets, modeling hydrological processes across a range of spatial and temporal scales, and predictability.

The improved availability of data holds the promise of improved predictions, regardless of the concerns about understanding small-scale processes raised above. Workshop deliberations pointed to research aimed at determining how the data can be best utilized to maximize the predictability from models. Data assimilation, where observations are merged with models, is well developed in the meteorology community. Research into data assimilation in the other areas of hydrologic and environmental sciences may be used to demonstrate how models can have synergy with measurements and to evaluate the predictability benefits from such approaches.

Multi-agency joint projects need to be devised to maximize the return for the resource investment and to engage a larger cross-section of the research and user communities. The fundamental issues regarding predictability and predictions are not restricted to a few variables such as precipitation or air temperature, but are pervasive across hydrologic and environmental sciences. One of the important issues identified both at the workshop and in numerous previous NRC reports is the need to reverse the degradation of existing monitoring systems where it can be demonstrated that the collection of consistent measurements and observations can lead to improved predictions of operational importance.

The key to success in research programs on predictability in hydrologic systems and in operational prediction programs is to develop strong linkages between research institutions and operational activities. Neither can fully realize their potential without recognizing their mutual synergies.

The workshop findings are consistent with those from an earlier COHS report on USGCRP (NRC, 1999) and the USGCRP Water Cycle Initiative Science Plan (Hornberger et al., 2001). These reports collectively define needed research that potentially has wide-spread and deep impacts on society by the incorporation of improved understanding of predictability into the operational arena.

In conclusion, discussions during the workshop and written contributions by the participants resulted in the definition of five research challenges and associated milestones, as presented in the previous section, that mark the path towards progress in advancing predictability research and understanding limits-to-prediction in the hydrologic sciences. The definition of such milestones is valuable for a research initiative because these milestones describe potential priority areas for research. More importantly, they define, in specific terms, where the community wants to see itself at different times along this path. Such milestones also provide benchmarks for tracking the maturation of a research area by identifying a vision for advancement and ensuring that the community has metrics for progress.

References

Hong, S-Y., and E. Kalnay. 2000. Role of sea-surface temperature and soil-moisture feedback in the 1998 Oklahoma-Texas drought. Nature 408: 842-844.

Hornberger, G. M., J. D. Aber, J. Bahr, R. C. Bales, K. Beven, E. Foufoula-Georgiou, G. Katul, J. L. Kinter III, R. D. Koster, D. P. Lettenmaier, D. McKnight, K. Miller, K. Mitchell, J. O. Roads, B. R. Scanlon, and E. Smith. 2001. A Plan for a New Science Initiative on the Global Water Cycle. U.S. Global Change Research Program, Washington, D.C.

NRC. 1999. Hydrologic Science Priorities for the U.S. Global Change Research Program: An Initial Assessment. National Academy Press, Washington, D.C.

NRC. 2001a. Basic Research Opportunities in Earth Science. National Academy Press, Washington, D.C.

NRC. 2001b. Grand Challenges in Environmental Sciences. National Academy Press, Washington, D.C.

Appendix A

Workshop Agenda and List of Participants

Committee on Hydrologic Science (COHS)
Workshop on
Predictability and Limits-to-Prediction for Hydrologic Systems

Damon Room
National Center for Atmospheric Research
1850 Table Mesa Drive
Boulder, Colorado 80305

<u>Thursday September 21, 2000</u>

8:30 a.m.	Breakfast available in the meeting room
9:15 a.m.	Introduction and strategy for workshop *Dara Entekhabi*, MIT
9:30 a.m.	Research and operations requirements in federal agencies *Panel of agency representatives*
9:45 a.m.	Predictability of regional hydrologic systems associated with terrestrial coupling *Randy Koster*, NASA Goddard Space Flight Center
10:10 a.m.	Observation-based predictability measures *Upmanu Lall*, Utah State University
10:35 a.m.	Break
11:00 a.m.	Hydrologic initialization and forecast in Numerical Weather Prediction *Dag Lohmann*, NCEP
11:25 a.m.	Operational seasonal prediction of hydroclimate over the US *Huug van den Dool*, NCEP

11:50 a.m.   Predictions, Predictability and Decision Making
                    Roger Pielke, Jr.

12:15 p.m.   Lunch at NCAR Cafeteria

1:15 p.m.   First panel discussion of science questions
                    Chair: *Marc Parlange*, Johns Hopkins
                    University
                        *1.   Are there predictable aspects of
                    terrestrial hydrology than can enhance
                    atmospheric weather and climate predictability?*

                        *2.   What are the stability and feedback
                    characteristics of two-way coupled subsurface,
                    surface, and atmospheric hydrologic systems?
                    How do they impact predictability?*

1:55 p.m.   Measures of predictability and effects of scale
                    on limits-to-prediction
                    Vijay Gupta, University of Colorado

2:20 p.m.   Characterizing and managing uncertainty
                    propagation in models
                    Roger Ghanem, Johns Hopkins University

2:45 p.m.   Defining measures for predictability and limit-
                    of-prediction in hydrology
                    Adam Schlosser, Center for Ocean-Land-
                    Atmosphere Studies

3:10 p.m.   Break

3:35 p.m.   Ensembles and predictability in climate and
                    hydrologic systems
                    Joe Tribbia, NCAR

4:00 p.m.	Second panel discussion of science questions Chair: Christa Peters-Lidard, Georgia Institute of Technology *3. What are the conceptual and model frameworks required to define limits-to-prediction in hydrologic systems?* *4. What are the data and records requirements to estimate the inherent limits-to-prediction directly from observations?*
4:40 p.m.	Adjourn

Friday September 22, 2000

8:30 a.m.	Breakfast available in the meeting room
9:30 a.m.	Emerging opportunities in predicting flood and flash-flood events *Baxter Vieux*, University of Oklahoma
9:55 a.m.	Current status and opportunities in quantitative precipitation estimation and forecast (QPE and QPF) *Witek Krajewski*, University of Iowa
10:20 a.m.	Extreme precipitation: Characterization of multiscale variability and predictability *Efi Foufoula-Georgiou*, University of Minnesota
10:45 a.m.	Break
11:05 a.m.	Predictability and limit-of-prediction in the global water cycle *Kevin Trenberth*, NCAR

11:30 a.m.	Third panel discussion of science questions

11:30 a.m.   Third panel discussion of science questions
               Chair: *Roni Avissar*, Rutgers University
               5. *What are the opportunities in extending
               the lead-time and accuracy of hydrologic
               predictions based on predictable weather and
               climate patterns so that they meet the
               requirements of water resource and other
               applications?*

               6. *What are the robustness and predict-
               ability criteria for models used in impact studies
               (e.g., hydrologic impacts of land use and global
               change)?*

12:10 p.m.   Lunch at NCAR Cafeteria

1:15 p.m.   Evolution of regional hydrologic and climate
               systems as an initial value problem
               Roger Pielke, Sr., Colorado State University

1:40 p.m.   Breakout groups to develop science plan and
               priorities for three sets of science questions;
               Collect and organize contributed bullet points

2:50 p.m.   Reconvene as group for general briefing and
               closing discussions

3:15 p.m.   Agency implementation of research and
               applications: Recommendations
               *Panel of agency representatives [NOAA (NWS,
               NCEP); NASA; USGS; U.S. Army Corps of
               Engineers]*

3:30 p.m.   Adjourn

4:00 - 6:30 p.m. NRC *Committee on Hydrologic Science* meets
                  to review workshop

Appendix B

Selected Papers Presented at Workshop on Predictability and Limits-to-Prediction In Hydrologic Systems

PREDICTABILITY OF REGIONAL HYDROLOGIC SYSTEMS ASSOCIATED WITH TERRESTRIAL COUPLING

Randy Koster
NASA Goddard Space Flight Center

Seasonal Prediction: Recent Results From NSIPP

Seasonal prediction of meteorological conditions cannot rely on the initialization and modeling of the atmosphere alone, since the timescales over which atmospheric anomalies dissipate are much too short. Seasonal forecasting must instead rely on the modeling of slower components of the earth system—namely, the oceans and the land surface. Although the ocean has the longer memory of the two, various studies (e.g., Kumar and Hoerling, 1995; Trenberth et al., 1998; Shukla, 1998; Koster et al., 2000) suggest that ocean conditions have only a limited impact on predictability over midlatitude continents. Thus, the memory associated with land surface soil moisture may turn out to be the chief source of midlatitude forecast skill.

The accurate initialization and modeling of soil moisture can contribute to a seasonal forecast only if two conditions are met: (1) the soil moisture has adequate "memory" (i.e., an anomaly lasts well into the forecast period) and (2) the atmosphere responds in a predictable way to the soil moisture anomaly. Various studies in the literature have addressed soil moisture memory and atmospheric response, both in the real world and in the modeling environment (Delworth and Manabe, 1988; Vinnikov et al., 1996; Huang et al., 1996; Liu and Avissar, 1999). In this paper, in place of a comprehensive literature review, we illustrate some key issues with recent research performed under the National Aeronautics and

43

Space Administration (NASA) Seasonal-to-Interannual Prediction Project (NSIPP).

Soil Moisture Memory

We recently manipulated the water balance equation at the soil surface into a relationship between the autocorrelation of soil moisture and the statistics of the atmospheric forcing, the variance of soil moisture at the beginning of the time period in question, and the structure of the land surface scheme used (Koster and Suarez, in prep.). The equation, despite its various approximations, successfully reproduces, to first order, the spatial distribution of soil moisture autocorrelation produced by the NSIPP modeling system. Figure 1, for example, shows that although many fine-scale details are missed, the equation captures the large-scale structure of the simulated 30-day-lagged autocorrelation for July. The equation works far better than the more traditional "water holding capacity divided by atmospheric demand" approach.

Further manipulation of the equation reveals four distinct physical controls on soil moisture memory: (1) temporal memory in the precipitation and radiation forcing fields, as perhaps induced by land–atmosphere feedback, (2) nonstationarity in the statistics of the forcing, as induced by seasonality, (3) reduction in anomaly size through the functional dependence of runoff on soil moisture, and (4) reduction in anomaly size through the functional dependence of evaporation on soil moisture. The relative importance of each control can be established through analysis of climate model data; thus, the equation can be used to characterize and explain geographical variations in simulated soil moisture memory. For example, the main physical control on memory loss in the eastern United States is seasonality of precipitation, and its impact is not large. Memory is reduced much more to the West because of the evaporation effect, which is influenced in part by low water holding capacities there. Autocorrelations in the far West increase again because of precipitation seasonality (acting in the opposite direction) and precipitation persistence.

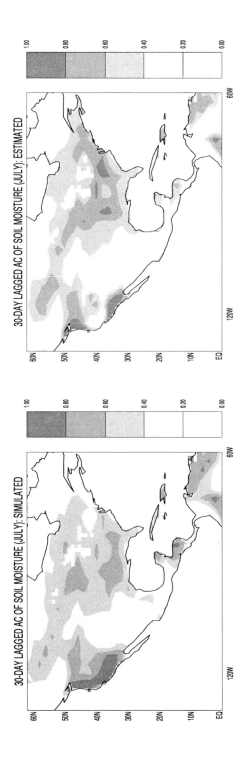

Atmospheric Response to Soil Moisture Anomalies

A recent study of the atmosphere's responsiveness to soil moisture anomalies focused on two ensembles of simulations with the NSIPP modeling system (Koster et al., 2000). Ensemble 1 consisted of 16 45-year simulations with interannually varying sea surface temperatures (SSTs) and interactive land surface processes. Ensemble 2 was similar except that land–atmosphere feedback was effectively deactivated; interannually varying land surface evaporation efficiencies (derived from a single member of Ensemble 1) were prescribed in each simulation of Ensemble 2.

Figure 2 shows the main result. The precipitation statistics from each ensemble were transformed into an index that describes the robustness of precipitation response to the specified boundary conditions. If, at a given point, all members of an ensemble produce basically the same time series of precipitation, then this index has a value close to 1, and we say that precipitation at that point is tied strongly to the surface boundary conditions—precipitation is predictable if the surface boundary conditions are themselves predictable (at least for the general circulation model (GCM) climate). If, on the other hand, the different ensemble members produce very different time series of precipitation, then the index is close to zero, and the potential for predictability is low. In this case, chaotic atmospheric dynamics overwhelm any control on precipitation imposed by the boundary conditions.

The left plot shows this "robustness" index over North America, as computed from boreal summer data (JJA) from Ensemble 1. Notice that foreknowledge of SSTs contributes to the predictability of precipitation only in the tropical areas. The right plot shows this index as computed from Ensemble 2. Foreknowledge of land surface moisture conditions leads to enhanced predictability over a significant part of midlatitude North America.

The land's contribution to precipitation predictability can be isolated by subtracting the values in the left plot from those in the right plot. Over North America, and in fact across the globe, the land contributions are highest in the transition zones between humid and dry areas.

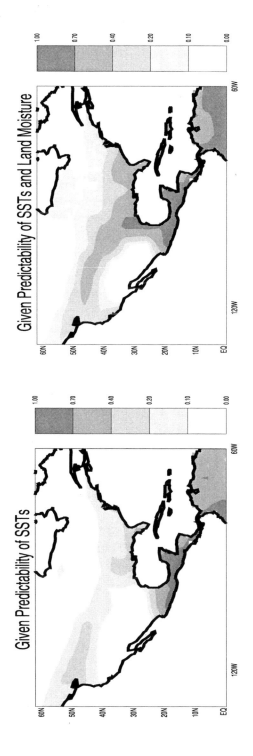

The low contribution in dry areas appears to reflect the low flux of evaporated moisture into the lower atmosphere. Contributions are low in humid areas partly because evaporation there is controlled more by atmospheric demand than by variations in soil moisture.

Unfortunately, Figures 1 and 2 cannot easily be compared, since the underlying simulations were performed at different resolutions—Figure 1 is based on runs with the 2X2.5 GCM, whereas Figure 2 is based on runs with the 4X5 version, with a correspondingly different climatology. Nevertheless, the figures suggest that land contributions can be high where they need to be—namely in regions with significant soil moisture memory. Indeed, soil moisture memory is fostered by land–atmosphere feedback that promotes precipitation persistence.

Two more results, though preliminary, are included here. The first comes from an idealized experiment in which all surface boundary conditions, including temperatures, are assumed to be perfectly known into the future. The NSIPP atmospheric GCM was first run for a specific July, using climatological SSTs. At each time step in the simulation, the values of all land surface model prognostic variables were written out to a special file. Then, an ensemble of 16 Julys using the same SSTs was run. At each time step of each member simulation, the updated values of all land surface prognostic variables were discarded and replaced by values read in from the special file. Thus, although the members of the ensemble differed because of their different atmospheric initial conditions, each was forced to maintain the same time series of (geographically varying) land surface prognostic variables. By quantifying the variations of atmospheric variables (precipitation, air temperature, etc.) seen between the ensemble members, using techniques similar to those used to generate Figure 2, we generate the estimates of land–atmosphere feedback strength shown in Figure 3a.

Note that this experiment is basically a simple, computationally cheap version of that which produced Figure 2. The idea is to promote an intercomparison of coupling strength among different models. Shown in Figures 3b and 3c are corresponding results for two other GCMs (Andrea Hahmann and Paul Dirmeyer,

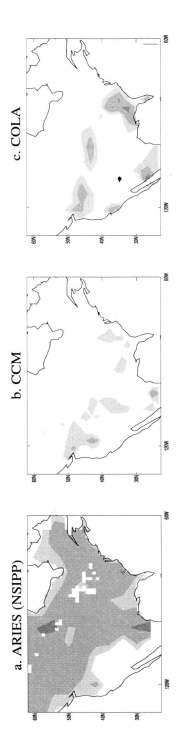

personal communication., 2000). The distinction between the GCMs is obvious; the NSIPP general circulation model (GCM) clearly shows a higher land–atmosphere feedback strength than either the CCM or COLA GCMs.

How do we know which level of feedback strength is most realistic? An additional experiment addresses this, though the results are currently inconclusive. The "control" in this experiment is an ensemble of Atmospheric Model Intercomparison Project (AMIP)-type simulations in which prescribed, realistic SSTs are used to force the GCM over the time period 1996–1999. The corresponding "experiment" ensemble is identical to the control ensemble except for one thing—at every time step in a member simulation, the precipitation generated by the GCM over the United States is replaced by observed precipitation (from a special hourly dataset generated by Wayne Higgins at National Centers for Environment Prediction (NCEP)) just before it hits the ground. Only the land surface feels this more realistic precipitation; the GCM's water vapor fields and the latent heating of the atmosphere are not replaced. The land surface presumably develops more realistic soil moisture states in response to the more realistic precipitation forcing.

Three global precipitation datasets are then compared: (1) the observed precipitation; (2) the precipitation from the AMIP-style runs (i.e., precipitation guided only by SST variability), and (3) the (constantly replaced) precipitation generated by the GCM in the experiment ensemble (i.e., precipitation guided by both SST variability and the presumably more realistic soil moistures). If the precipitation generated in the experiment ensemble is significantly closer to the observations than that generated in the AMIP ensemble, then we will have demonstrated a positive impact of more realistic soil moisture on precipitation in the GCM, and we will have also shown that land–atmosphere feedback is operating in the real world. Some improvement is indeed seen in our preliminary runs—Figure 4 shows significant reductions in precipitation error over the United States, especially in summer.

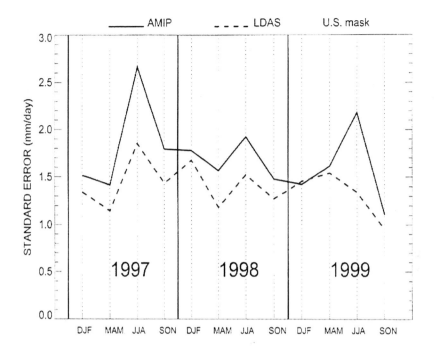

FIGURE 4

Seasonal Prediction:
Science Questions and Research Needs

The results above serve as background for three science questions related to seasonal prediction.

1. *Over what timescales can soil moisture be predicted in the real world? How well can these timescales be simulated in a modeling system?* The analysis leading to Figure 1 shows that several factors can influence soil moisture memory. How relevant is each factor in the real world, and do GCMs simulate their relative importance correctly? Can we specify regions and seasons for which useful soil moisture memory is essentially unattainable?

2. *To what degree is the atmosphere "coupled"' to the land surface in the real world? How does the degree of coupling differ among modeling systems, and how do we evaluate the realism of this coupling?* The differences seen between modeling systems in Figure 3 are significant and disturbing. Because modeling systems are key to long-term hydrological prediction, differences in their coupling strength must be quantified, understood, and evaluated against observations.

3. *Are some hydrologic states more prone to long-term memory than others? Why?* The analysis of soil moisture memory (beyond what is presented above) suggests that some hydrologic anomalies, such as drought, may have potential for added persistence. In addition, some meteorological regimes may be more conducive to land–atmosphere feedback and thus may promote persistence. If we find that predictability in a region is generally poor, can we at least hope to predict the occurrence and/or longevity of certain extremes in the region?

Some New and Relevant Initiatives

Although quantifying land–atmosphere feedback strength from a strictly observational analysis is very difficult, a combination of modeling and observations can provide essential insight. Two scientific panels, namely the Global Energy and Water Cycle Experiment (GEWEX) Global Land-Atmosphere System Study (GLASS) panel and the U.S. Climate Variability and Predictability (CLIVAR) Seasonal-to-Interannual Modeling and Prediction (SIMAP) panel, are now promoting a model intercomparison experiment similar to the one that produced Figure 4. The experiment, which would be global in scope and would cover additional years, would have two phases. The first phase, the "maintained soil moisture" phase, is designed to quantify the impact of realistic soil moisture contents on precipitation. In this phase, the coupled land–atmosphere model is forced to maintain realistic soil moistures at all times, so that the focus is mainly on atmospheric response to land conditions (i.e., on the strength of the coupling). The second phase, the "initialization" phase, is similar except that the artificial replacement of precipitation is halted at some prearranged time. The second phase specifically addresses persistence, being designed to establish the impact of a realistic land surface moisture initialization on precipitation forecasts.

The first phase should address the second question above regarding the degree to which the atmosphere is coupled to the land surface.. An improvement in simulated precipitation due to more realistic soil moistures, if it occurs, would serve both to demonstrate the existence of coupling in the real world and to validate the coupling strength in the modeling system. The second phase, in which soil moistures are initialized at realistic values and then allowed to evolve, will help address the first question regarding the modeling of soil moisture timescales and will, more generally, quantify predictability in the participating models.

Such a modeling effort should be supplemented, of course, by observational studies. The initialization phase of the above experiment could be improved through application of superior initial conditions, as obtained through a land data assimilation system (LDAS) approach. Also, observational analysis of soil moisture and meteorological data outside the modeling context will help establish the real world's prevailing timescales. Currently, such observational analyses are limited by the paucity of historical data; efforts to expand and maintain the observational database are needed.

The modeling experiment that led to Figure 3 is also being promoted by the two science panels, though to a lesser degree. The experiment is highly idealized but at least has the advantage of being easy and computationally inexpensive to perform. It thus can shed light quickly on intermodel differences in coupling.

Role of Federal Agencies

Demonstrating the effectiveness of earth science models for hydrologic prediction, either through the experiment described in 2.2 above or through some other well-structured model intercomparison study, requires substantial computer resources, particularly because such studies will require ensembles of long-term simulations to produce the data needed for statistical analysis. Funding agencies should be ready to support these needs. In addition, maintenance and continuation of observational programs, particularly for soil moisture and meteorological data, are critical for providing a basis for the evaluation of prediction systems.

HYDROLOGIC INITIALIZATION AND FORECAST
IN NUMERICAL WEATHER PREDICTION

Dag Lohmann
National Centers for Environmental Prediction (NCEP)

Uncertainty and systematic errors in the prediction of hydrologic states and fluxes, such as soil moisture, evapotranspiration, and runoff, are introduced into the forecast system by three different factors. This paper will concentrate on the third factor and how it is addressed in numerical weather prediction, but will mention the other two as well. The modeling strategy is highlighted in an NCEP-modified "Shukla staircase" (see GAPP science plan), with the model hierarchy on the outside, covering the various spatial scales, and the data assimilation system in the inside, for determining the proper initial conditions for the models.

Hydrologic Model Physics and Problems
in Parameter Estimation

Once the physical knowledge about the system is expressed in a mathematical model, one has to find optimal parameters for the model. Hydrological models use many different equations for their model formulation. The hydrologic models, which are currently coupled to atmospheric models, are rather simplistic in the description of horizontal and vertical water fluxes in the soil, while they often have a detailed description of vegetation processes. The reason is that most of these models originate from atmospheric modelers, who traditionally place a greater emphasis on the energy side of hydrologic processes (turbulent fluxes and ground heat flux), while hydrologists are more concerned with the water balance, especially runoff and streamflow. This led to the development of many based on Richards equation 1-dimensional land-surface models (LSMs), which are used on the scale where hydrologists would use conceptual lumped models like the Sacramento model. Only a few LSMs have incorporated "hydrologic knowledge," which makes their use for hydrological predictions promising. All LSMs and traditional hydrologic models have parameters, such as hydraulic conductivity and root depth, which can only be inferred indirectly from modeling exercises or interpolated from sparse point measurements.

Resolution of the Atmospheric and the Hydrological Models

Doubling the horizontal resolution of atmospheric models (such as the NCEP estimated time of arrival (ETA) model) increases the execution time roughly by a factor of eight. It therefore will take some time before atmospheric models have a horizontal resolution of well known "as physical as possible" hydrologic models; (on the order of 100 meters). Numerical weather prediction can be shown to benefit from both higher resolution and a physics upgrade for example the coupled LSM.

Uncertainty in the Initialization of the Models

State variables in many of the models have to be initialized. This has led to the development of sophisticated variational methods in atmospheric and ocean sciences, while for operational purposes, methods to initialize land surface models are still in the early stages of development. The soil moisture data from the 4-dimensional data assimilation systems (4DDA) of coupled land–atmosphere models often suffer substantial errors and drift owing to precipitation, temperature, and radiation biases in the land-surface forcing of the coupled system. To constrain such errors and drift, some developers apply soil moisture nudging techniques in coupled 4DDA, but such nudging can introduce other undesirable behavior such as over amplified annual cycles of soil moisture and lack of water conservation.

As an appealing alternative to coupled land-surface 4DDA, a consortium of Continental-Scale International Experiment (GCIP)-supported groups has undertaken the development and execution of an uncoupled Land Data Assimilation System (LDAS). The LDAS execution is hosted on an NCEP developmental computing platform and is supported by NCEP in collaboration with NASA/ Goddard Space Flight Center (GSFC), National Weather Service Office of Hydrology, National Environmental Satellite, Data, and Information Service/Office of Research and applications (NESDIS/ORA), Princeton and Rutgers Universities, the University of Washington and the University of Maryland.

Specifically, these partners are developing, executing, and validating a prototype national, realtime, hourly, 1/8-th degree, distributed, uncoupled, land-surface simulation system. This system consists of several land-surface models (LSMs) running in tandem on a common grid and driven by common surface forcing fields. The hallmarks of the forcing fields are observed, hourly, gage/radar precipitation and observed Geostationary Operational Environmental Satellite (GOES)-based satellite-derived surface solar insolation. Additionally, a common streamflow routing model is being applied to each LSM's gridded runoff on the shared common grid. Finally, the "DA" for "data assimilation" in LDAS denotes a later "Phase II" thrust that will include the assimilation of satellite-derived land-surface fields, such as skin temperature, soil moisture, snowpack, and vegetation density and greenness. Another future but central LDAS phase will be a forecast component, wherein LDAS will be integrated days, weeks, and months into the future using ensemble surface forcing (including ensemble quantitative precipitation forecasting (QPF)) from the weather and seasonal prediction models of NCEP.

With this setup, which is currently under construction, it is hoped that the state variables will be initialized as well as possible; first results look promising.

A number of important questions and discussion points emerge from the preceeding text.

Priority science questions in predictability of hydrologic systems:

- How do we estimate model parameters in hydrologic models?
- What is the required complexity / structure for hydrologic models?

Required modeling and analysis initiatives and data collection programs:

- There is not enough analysis done with the data already produced.
- More cross-validation is needed to decide about needed model complexity.

Roles of federal agencies that fund research and support monitoring:

- Federal agencies should be funded to implement operative test-beds for research. This would enable researches of the academic world and federal and state agencies to bridge the gap between research and day-to-day operations.

OPERATIONAL SEASONAL PREDICTION OF HYDROCLIMATE OVER THE UNITED STATES

Huug van den Dool
National Centers for Environmental Prediction

INTRODUCTION

The Climate Prediction Center (CPC/NCEP) has developed a physically based soil moisture data set for the United States for the period 1932 to the present. This data set is the result of integrating the so-called "leaky" bucket of Huang et al. (1996) for nearly 70 years with observed forcing, most notably observed precipitation at 344 Climate Division, but also temperature. The soil moisture data set, with daily updates through yesterday 12Z, thanks to daily precipitation analysis by Higgins and Shi, has been widely applied for drought and wetness monitoring, the launching of monthly/seasonal prediction tools, and many research projects. Other outputs of the model include evaporation, runoff, etc., for all data from 1932 to the present (lower 48 states only), allowing a judgment of present conditions in historical perspective, forming anomalies, etc.

CPC is the organization that issues the official monthly and seasonal forecasts for the U.S. Traditionally, the predictands were time averaged temperature (*T*) and precipitation (*P*). Recently, antecedent soil moisture conditions have been used in several emerging systematic methods to improve T and P forecasts, especially in summer. Experimental forecasts for soil moisture itself are also emerging and the possibility of some kind of probabilistic monthly/seasonal runoff prediction is being considered jointly with Office of Hydrology/National Weather Service (OH/NWS).

Note that in this paper, we are not discussing major NCEP activities in data assimilation for the shorter-range forecasts. The soil plays an important role in these shorter-range forecasts, one main problem being drift of the data assimilation system due to biased precipitation input. We will only mention below the LDAS activity, which is a modern data assimilation, but observed precipitation is used to avoid such drifts.

As time permits we will present/discuss some or all of the following:

• Monitoring of the current situation. The United States as a whole is and has been very dry. This will be discussed by looking at soil moisture anomalies as of mid-September 2000. Recent changes will be presented as well.

• Incorporation of short-term forecasts. Taking the medium-range forecast (MRF) ensembles at face value, the bucket model can be integrated 1 and 2 weeks ahead for short-term forecasts.

• The most recent soil moisture anomalies, after recasting the units where necessary, are used as the initial lower boundary condition in NCEP's ensemble of "coupled model" runs for multiseasonal forecasts.

• Likewise, the most recent soil moisture anomalies are used to launch empirical forecast tools. An example *C*onstructed *A*nalogue on *S*oil (CAS) moisture method for JJA2000. The CAS method can also be used for process studies on academic initial conditions in soil moisture.

• The aftermath of a La Niña in 2000. This too was a consideration in forecasts made in early 2000. Some La Niña composites of soil moisture will be shown.

• Some comments on empirical evidence of interactive soil moisture over the United States. Further analysis: comparing model runs with and without interactive soil moisture, in terms of tempera-ture persistence, lagged precipitation–temperature correlation, etc.

• The latest Atmospheric GCM used at NCEP for ocean–land "coupled" runs. Analyses of monthly mean temperature, soil moisture, evaporation, and precipitation in 11 Atmospheric Model Intercomparison Project (AMIP) global SST forced runs by this model reveal both strengths and weaknesses. We briefly discuss (for the United States only) the mean state, the interannual variability, spatial degrees of freedom, and various lagged correlation, prediction skill (model vs. observation) and predictability (model vs. model).

The Huang et al. (1996) model has also been run for all other continents, resulting in a global soil moisture data set. This data set for 1979–1999 is updated only about once a year, not as religiously as the U.S. data set, which has operational application. The global data sets are mainly used for two kinds of applications: one where soil moisture is a prescribed lower boundary condition, and the other for various geodetic research activities such as investigation of the feasibility of detecting soil moisture anomalies by measuring gravity from satellite and determinations of the annual cycle of the geoid etc.

Near Future (2-3 years)

The Environmental Modeling Center (EMC) of the NCEP, as a participant in the multi-agency GCIP Land Data Assimilation System (LDAS) project, has started to run its model, the NOAH LSM, in stand-alone mode with observed forcing functions including observed rainfall, from spring 1999 forward. The NOAH model is physically far more complete than the CPC leaky bucket, and it has much higher spatial (1/8th of a degree) and temporal (hourly) resolution. By using observed forcing, one avoids some of the usual biases of coupled data assimilation. In this sense, the philosophy is the same as Huang et al. (1996) model, but LDAS is far more comprehensive. The LDAS data set (currently no more than 2 years) would be extremely useful for climate application if it covered a multidecadal period. We think it is feasible to integrate the NOAH LSM from 1948 to the present. A 1998 pilot, in concert with Regional Reanalysis, is being prepared. This would require the assembly of many forcing data sets at high resolution for a very long period. The pay-off of this endeavor would be (1) improved soil moisture data for the various CPC prediction and monitoring tools as well as for the community, (2)superior model consistent initial conditions for numerical prediction on time scales ranging from hours to seasons, and (3) model consistent prescribed lower boundary conditions in AMIP-like experiments to determine predictability due to soil moisture anomalies.

Priority science questions in predictability are first and foremost the design of experiments that could give us pertinent information about hydrology and atmospheric predictability. Which one is first—the atmosphere or the soil? If the soil itself is the prediction target, much depends on the precipitation forecast (a very weak link in the chain). If the atmosphere is the forecast target, the characterization of the lower boundary down to some depth is all-important. Taking the latter point of view, issues are the mixing of SST (global or otherwise) and initial soil moisture, the challenging formulation of lower boundary conditions vs. initial value predictability (is soil moisture the initial value?), and the question of what we can learn from flawed models (which have overactive land–atmosphere interaction). We advocate the design of data studies as a sanity check on numerical models.

In physics, observations are normally all-important. Note that the land surface hydrology is plagued by a lack of data. In spite of its name, LDAS does not assimilate any data. Precipitation observations are difficult, precipitation analysis is a challenge, (especially near orography), soil moisture observations scarce and problematic, and evaporation observation (over any length of time) are almost nonexistent. We need to take a comprehensive look at what data we have, what we ideally want to try to accomplish, and how further observations would help.

PREDICTABILITY OF HYDROLOGIC SYSTEMS IN THE CONTEXT OF A NEW INTERDISCIPLINARY FRAMEWORK LINKING WATER, EARTH, BIOTA (WEB)

Vijay K. Gupta
University of Colorado, Boulder

WEB Vision

Water, Earth, Biota (WEB) is a new interdisciplinary research and education framework that addresses the central role of water in linking landscapes, atmosphere and oceans, geochemistry, and biota in spatial scales from the molecular to the planetary, and in temporal scales from instantaneous to geologic. The perspective of interconnectedness at multiple scales is fundamental for gaining a holistic understanding of the complexity of nature and of changes over space and time through interactions with humans. Increasing threats to natural environments make this broad vision critical for the management of water and other natural resources to sustain growing human populations and modernizing economies. The WEB vision does not stop with the importance of the topic and the definition of science goals; it identifies a path for moving forward.

Key Issues Driving the WEB Imperative

Key issues driving the WEB imperative include the following:

- Water is the life-blood of the planet. The water cycle comes together at the planetary scale and must be studied as a planetary process to gain holistic understanding.
- Regional interests, agency missions, disciplinary training, and the institutionalization of field experiments have intellectually partitioned water studies into domains.
- Scientists working in each domain increasingly recognize that their findings depend highly on boundary conditions at the interfaces with the other domains.
- As demands on a fixed quantity of water increase and adversely impact the environ-ment, water managers are driven to greater efforts to contain floods, droughts, and contamination while being increasingly stymied by an inadequate science framework for quantifying feedbacks among the domains.
- Research is needed to assess and extend the predictability of a wide diversity of interconnected hydrologic systems within the broad framework of WEB at many scales to allow better risk and vulnerability assessments that support resource management decisions.
- Progress toward this WEB ideal will require an observation system that compiles and freely distributes coordinated data on WEB-related processes at a variety of space and time scales. Research and measurement must be planned together.
- Continued progress will also require an interdisciplinary initiative for education and public outreach that will enable students and the general public to think about water issues from both planetary and local perspectives.

Societal Rationale

Our limited understanding of major couplings of water, energy, and biota leaves policy makers poorly positioned to guide natural resource development and management to serve growing human populations while sustaining environmental quality. Stakes are high, as society is making critical decisions on water supply, floods, and pollution from a fragmented understanding that could prove to be quite dangerous over time.

Science Challenges

The WEB framework identifies four main methodological steps—scaling, coupling, diagnosing, and modeling—to reach consistent conclusions in the face of space–time variability and dynamic nonlinearities that confound cross-scale computations and understanding. These methodological steps are a key to advancing predictability and to understanding the limits-to-prediction of a wide diversity of interconnected hydrologic systems. The WEB framework identifies four separate domains—water cycling, energy, earth structure, and ecology—and the need to understand the ties among them. The WEB report (http://cires.colorado.edu/hydrology) lists multiple scales for research. Each scale has its own level of complexity. A key scientific challenge is to be able to cascade across scales and levels of complexity in a self-consistent manner. Some examples are offered to explain what this means at the global scale, continental scale, drainage basin/aquifer scale, and hillslope scale. For illustration, we list below an example of a science challenge at the global scale. Examples of sciences challenges at other scales are given in the full WEB report.

Example of a Science Challenge at the Global Scale

Global climate varies over time as a result of complex, nonlinear couplings involving the water cycle in ocean, atmosphere, and land. The maintenance of a stable environment by the world's biota is perhaps the most significant phenomenon in ecology (Reiners, 1988). Atmospheric water vapor and the conflicting effects of clouds on radiation (Webster, 1994) are linked to the hydrologic cycle via cloud formation, precipitation, and evapotranspiration. We are far from a quantitative understanding of (1) the role of the water cycle and biota in critical relationships that govern natural quasi-oscillatory climate variability at interannual, interdecadal, and longer time scales, (2) the combined effects of these interactions, and (3) the robustness of world climate to anthropogenic impacts. Large-scale organized components of the hydrologic cycle, such as monsoonal systems, significantly influence seasonal and longer variations in vegetation and may in turn be influenced by large-scale land use changes. These issues need to be investigated to asses the predictability, risks, and vulnerability of the hydrologic cycle in the context of the hypothesis that global climate is a self-regulating emergent phenomena due to nonlinear coupling between biota and the rest of the earth's climate system (Lovelock, 1995).

WEB in River Basins as an Illustrative Theme

River basins integrate physical, chemical, and biological processes in the spatial and temporal organizations of fluxes and structures. Science issues that require focused data collection, hypothesis development, and tests and that have the potential for significant cross-disciplinary theoretical advances are identified in a river-basin context as an illustrative example. A number of other settings should be similarly examined. For the river-basin example, major contributions to science could be made through a better understanding of (1) earth-related processes, (2) atmosphere-related processes, (3) climate-related processes, (4) geochemical processes, and (5) ecological processes. It is argued that we need to focus on the dynamics of how these processes integrate in the evolution of planetary structure and life environment.

Suggestive Hypotheses as Examples of Research Challenges

The following list presents some of the working hypotheses, identified by the WEB activities which provide an important framework for future research.

• The river basin is a self-organizing system. The space/time distributions of water, chemicals, and biota guide the evolution of land structure, drainage networks, habitats, and life itself.

• River basin-scale processes modulate local processes and thereby alter weather, geochemical fluxes, habitats, and biodiversity—impacts to be captured in downscaling.

• River basin-scale processes also impact global climate by the way they partition land–atmosphere–ocean fluxes and recharge to/discharge from aquifers—impacts to be captured in upscaling.

• The water and energy fluxes through ocean, atmosphere, and land at multiple scales in space and time interact through strongly nonlinear connections. These connections must be defined and quantified to simulate planetary environmental dynamics.

• Human modification to a river basin alters local structure and fluxes in ways that have significant impacts on both upscaling and downscaling.

An Interdisciplinary Initiative for Education
and Public Outreach

An interdisciplinary initiative for education and public outreach will enable students and the general public to understand the unique role of the planetary water cycle in the co-evolution of life and of a habitable climate on Earth for 3.8 billion years and also to understand and solve the water problems facing society.

WEB Infrastructure

We propose moving ahead by establishing "Natural Laboratories" and a "National Hydrologic Facility" to organize and analyze comprehensive data sets, both existing and new. We are seeking an administrative mechanism to facilitate the deployment of instrumentation; to coordinate data collection, archiving, and access for integrative large-scale research; and to facilitate education and technology transfer. A multistage approach is necessary to develop infrastructure over time. The procedures will take continuing discussion, and we set forth here a strawman to get thinking going. Our idea is for Phase I to be a pilot project focused on a science theme at the river-basin scale. A WEB Office for Science Support (WOSS) would be established to reach out to additional themes and to facilitate mechanisms for cross-coordination. Some generic issues related to the selection and management of a network of natural laboratories are described in the full WEB report, and a brief description of Phase II is given.

IDENTIFYING AND EVALUATING PREDICTABILITY FOR LAND–ATMOSPHERE AND HYDROLOGIC PREDICTION SYSTEMS

C. Adam Schlosser
NASA Goddard Space Flight Center

Where's the Predictability?

Historically, the term "predictability" has been separated into two categories (e.g., Lorenz, 1975) and has been applied to the framework of ensemble predictions (of a physical system that is inherently chaotic). What is commonly referred to as the "initial value problem" is one aspect of predictability. Given hypothetically "perfect" information with which to initialize an ensemble forecast, how can we determine how long that "perfect" information will persist into the forecast in the presence of propagating errors that exist in the prediction system? The other aspect of predictability is associated with an externally forced response that is, to a quantitative degree, ubiquitous among the forecast members of an ensemble prediction (e.g., all forecast members predict consistent soil-water variations in response to a precipitation anomaly).

To identify predictability within "coupled land–atmosphere prediction systems" (e.g., operational weather forecast models and global climate models), previous research has aimed to show which elements of the land–surface system are central to this coupled mechanism. It has been exhaustively shown that soil moisture, and its persistence, play a critical role in coupled land–atmosphere variations. Generally speaking, however, it is the liquid and frozen storage of water on the ground (i.e., snow cover and interception storage) and in the ground (i.e., soil water that is available for exchange processes between the land–surface and overlying atmosphere) that are central to coupled land–atmosphere variations. These coupled variations result from the fact that the aforementioned terrestrial water storages have a persistence time-scale that is comparable to or longer than the characteristic time-scales of the interacting atmospheric processes. Therefore, it follows that regions of strong persistence of "active" water storage (i.e., terrestrial water storage available to exchange processes between the land and the atmosphere) will lead to predictability of these water storages and potentially impact atmospheric predictability (as shown by Schlosser and Milly, 2000), which can in turn lead to enhanced land predictability (i.e., predictability via a forced response).

An important additional source of predictability that is external to local land–atmosphere coupling is the impact of sea surface temperature (SST) variations on remote responses of the atmosphere. Certainly, if a given SST pattern (e.g., El Niño) can impose a forced, and predictable, response of the atmosphere over a continental region (e.g., a precipitation anomaly), this could, in turn, produce a predictable signal in the land-surface (i.e., predictability of water storage). However, the ability to which the modified land-surface then imposes a predictable effect on the local land-atmosphere coupling will be largely dependent on land-surface persistence. In other words, when the remotely forced SST response of the atmosphere abates, to what degree will the resulting land anomaly impact subsequent hydroclimatic variability? Here then, we return to the importance of continental water–storage persistence and of its impact on coupled land–atmosphere variability and predictability.

With regard to "hydrologic prediction" (i.e., streamflow, river discharge, and flood/ drought prediction), the challenge lies in translating the predictable information of a (presumably) fully coupled ocean–land–atmosphere (i.e., climate/weather) prediction system (which, in practice, has not explicitly incorporated hydrologic networks) to hydrologic prediction/operational systems, and in assuring that those quantities that are relevant to the information transfer are predictable. For example, for a river–routing scheme to produce a predictable hydrograph, the scheme should receive surface and subsurface runoff rates (which could be provided by gridded land–surface outputs of the climate prediction system) that are predictable. This requires that to a large degree, the relevant simulated quantities of the ocean–land–atmosphere prediction system (i.e., precipitation, soil moisture) are predictable in order to produce predictable runoff rates. Even still, we must verify that the resulting runoff rates provided to the hydrologic prediction system are predictable.

Is the Predictability for Real?

Continuing with our discussion of predictability through persistence, if we were to find a region within a modeled climate that contains a strong potential for predictability (i.e., strong persistence), can we be assured that the persistence of the model is realistic and therefore is, in a sense, reflecting a "true" potential for predictability? For example, if a persistent (and therefore predictable) anomaly of soil moisture then creates a predictable response of the atmosphere (e.g., a corresponding precipitation and/or temperature anomaly), can we be assured that these predictable signals are real within the coupled model prediction system? In order to answer this question in the context of coupled land–atmosphere modes of predictability, it is necessary to identify whether or not the relevant components of the prediction system (e.g., the soil moisture scheme and/or the precipitation scheme) are behaving in a manner that is consistent to realistic/ observed modes of variability.

To that end, we should identify and verify the sources and controls of the persistence (and therefore the predictability) of terrestrial water storage. The landmark study by Delworth and Manabe (1988) presented a direct relation between simulated soil moisture persistence in a "bucket hydrology" scheme to a first-order Markov process. This direct comparison was possible due, in part, to the linear relationship between soil moisture and evaporative stress in the bucket model (among other simplifying assumptions). However, the more complex "big leaf" models that have been developed within the past decade contain highly nonlinear relations between soil water storage and runoff and evaporation processes. As a result, an analytical description of the controls of soil moisture persistence becomes more complex. However, a recent study by Koster and Milly (1997) was able to effectively explain the disparity of the simulated annual water budgets of the participating models of the Project for the Intercomparison of Landsurface Parameterizations Schemes (PILPS) through linear approximations of evaporation and runoff controls as a function of soil moisture. The results underscore the potential to effectively summarize and verify the degree and controls of soil-water persistence for the variety of models used to represent the continental surfaces in weather and climate models. More importantly however, we must apply these techniques with corresponding observed quantities in order to verify that our model-based assessments of predictability are reflective of what would (hypothetically) occur in nature.

Similarly for hydrologic prediction, it is crucial that we evaluate the sources of predictability that contribute to the predictability of a hydrologic prediction system. For example, if a climate model were able to provide a predictable forecast of total runoff rates for grid points that span a hydrologic basin, inputted into a river routing prediction system, we must be assured that not only the predictable signal of runoff is robust and realistic, but that also the partitioning of the runoff rates (between surface and subsurface flow) are realistic. Otherwise, the resulting hydrograph prediction, while being predictable, could result in low prediction skill due to the impact of inaccurate runoff partitioning on the resulting streamflow estimates. The impact of the disparity in runoff partitioning for streamflow prediction has been demonstrated in the PILPS 2c simulations by Lohmann et al. (1998).

Therefore, it is important that evaluation of current models used in hydroclimatological prediction are rigorously continued and that the diagnostics used in the evaluation not only aim to identify systematic biases/errors, but also to assess the impacts of the errors/biases on predictability. This will ensure that the predictability that is identified is realistic, and it will build our confidence in the predictive ability of our hydroclimatological prediction systems.

ENSEMBLES AND PREDICTABILITY IN CLIMATE AND HYDROLOGIC SYSTEMS

Joseph J. Tribbia
National Center for Atmospheric Research

Ensembles on the Synoptic Scale

The use of ensembles in weather forecasting is an idea going back to the early attempts to perform stochastic-dynamic forecasts. The moment prediction technique proposed by Epstein (1969) was seen to be computationally infeasible, so Leith (1974) proposed a Monte-Carlo method (i.e., a randomly perturbed ensemble of weather forecasts) in order to provide the two major benefits afforded in the stochastic-dynamic approach. The beneficial attributes to be gained are the improved skill (in a statistical sense) of the mean forecast, which is gained through the nonlinear filtering of the uncertain forecast degrees of freedom/ variables, and an estimate of the uncertainty or reliability of the forecast through the prediction of the variance/ covariance.

Ensemble prediction is now an everyday part of operational short- and medium-range weather forecasting at European Center for Medium Range Weather Forecasts (ECMWF) and National Centers for Environmental Prediction (NCEP), the major global weather centers. The professed goal of the ensemble prediction systems at both centers is more ambitious than that stated above; the aim is to predict the probabilities of event occurrences at forecast locations. This elevated goal is not merely desirable but partly necessitated by the requirement that ensemble predictions be objectively verifiable. Because the number of forecasts within an ensemble is proportional to computational costs, to date, ensemble size has been limited to at most 50 members. Because of these limitations, elaborate techniques of ensemble member initialization, using methods imported from dynamic systems theory, have been used to maximize the information content of the small-sized ensemble—small, that is, when compared to the phase space dimension of an operational forecast model (10^6). These methods of initialization are primarily designed to capture the most uncertain aspects of the dynamic meteorology within a forecast, with little or no concern for the uncertainty in the physical processes (e.g., moist processes) and the skill with which precipitation can be forecasted. Nonetheless, these techniques can often give reasonable estimates of the uncertainty of synoptic scale precipitation.

The situation is somewhat different on both the very short range (<24 hours) and long range (one month to a season). Because of the high sensitivity of precipitating mesoscale phenomena to variations in model formulation and low-level moisture, uncertainty in these aspects will likely require a different approach than the one taken on the synoptic scale. At long range, the initial conditions of an ensemble are not treated as elaborately as in the short and medium range, since at these time scales initial conditions are nearly completely forgotten, and anomalous boundary forcing is the sole residual memory driving predictability. The applications of ensemble methods in these two forecast problems, with particular attention to precipitation, are discussed individually below.

Mesoscale Ensemble Prediction

The use of ensemble methods for mesoscale prediction is in its infancy and should properly be considered a research problem. The goals of ensemble prediction on this scale include the prediction of probability of precipitation and the probability of extreme events. Recent research by Crook (1996) and Errico and Stensrud (2000) demonstrates the extreme sensitivity of predicted precipitation coverage and amount to variations in low-level moisture and to changes in convection parameterization. This latter uncertainty reflects both the real lack of scientific consensus and understanding in the current generation of convective parameterizations and the stochastic nature of the relationships between convection and the larger-scale environ-ment. This being the state of affairs on the mesoscale, the most rational methodology may be one in which ensembles include both variations in initial conditions and variations in formulations. Such a strategy might be the most accurate manner of providing threat probabilities, probability of precipitation, and ranges of precipitation amount.

Long-Range Ensemble Prediction

For extended range prediction, the use of ensemble methods is a necessity since weather systems, and characterizing their effect on the large-scale, low-frequency patterns for long-range prediction, are not predictable beyond two weeks. At the present time, there is still active research concerning the statistically optimal method of utilizing ensemble forecasting for long-range predictions. Operationally, at centers like the International Research Institute for Climate Predictions (IRI) and European Center for Medium Range Weather Forecasts, the main use of the ensemble is as a nonlinear filter to generate the most reproducible components of a forecast and to determine a signal-to-noise ratio from the ensemble mean and (ensemble) standard deviation. With respect to precipitation, global prediction and climate models have significant mean climate errors in both the location and magnitude of precipitation. Because the targeted forecast problem is the response to boundary anomalies, (frequently tropical ocean sea surface temperature anomalies), both the climate mean and the anomalous response are subject to a climate drift. The presence of systematic errors and climate drift once again suggests that the most rational strategy for ensemble prediction will be a multimodel, multi-initial condition ensemble method.

Concluding Remarks

Because the current generation of global models and mesoscale models have strong systematic biases in the monthly to seasonal timeframe for global models and 3- to 12-hour range for mesoscale models, the optimal ensemble strategies for precipitation prediction in these time and space scales will include a multi-model ensemble. There remains a great need to analyze and rectify the errors in current methods of parameterization of moist processes.

PEDICTABILITY OF FLASH FLOODS USING DISTRIBUTED PARAMETER PHYSICS-BASED MODELS

Baxter E. Vieux
University of Oklahoma

Why Distributed Parameter Modeling?

Historical climate modeling practice has been to use lumped representations of physical processes because of computational limitations or because sufficient data were not available to populate a distributed model database. How one represents the process in the mathematical analogy and implements it in the hydrologic model determines the degree to which we classify a model as lumped or distributed. Several distinctions on the degree of lumping can be made in order to better characterize a mathematical model, the parameters/input, and the model implementation.

The main advantage of distributed modeling is that the spatial variability of parameters and rainfall input is incorporated into the model response. Each parameter and rainfall input has its own spatial variability, which can be characterized by informational entropy (Vieux, 1993). Incorporating spatial variability of a particular parameter related to vegetative cover, soil infiltration rates, rainfall, or topographic characteristics may or may not have important conse-quences on the hydrograph response of the model. If the parameter has a narrow range of variation or if the parameter has only a few values represented in the basin, then its spatial variability may have little effect on simulated hydrographs.

Mathematical Analogy

A numeric solution of the governing equations in a physics-based model employs discrete elements. The three representative types are finite difference, finite element, and stream tubes. At the level of a computational element, a parameter is regarded as being representative of an average process. Thus, some average property is valid over the computational element used to represent the process of flow. For example, porosity is a property of the soil medium, but it has no meaning at the level of the pore space itself. Physics-based models solve governing equations derived from conservation of mass, momentum, and energy. Unlike empirically based models, differential equations are used to describe the flow of water over the land surface or through porous media to describe, or energy balance in the exchange of water vapor through evapotranspiration.

If the physical character of the hydrologic process is not supported by a particular analogy, then errors result in the physical representation. Difficulties also arise from the simplifications because the terms discarded may have afforded a complete solution while their absence causes mathematical discontinuities. This is particularly true in the kinematic wave analogy, in which changes in parameter values can cause discontinuities, sometimes referred to as *shock* in the equation solution. Special treatment is required to achieve solution to the kinematic wave analogy of runoff over a spatially variable surface. Vieux et al. (1990) and Vieux (1991) presented such a solution using nodal values of parameters in a finite element solution. This method effectively treats changes in parameter values by interpolating their values across finite elements. The advantage of this approach is that the kinematic wave analogy can be applied to a spatially variable surface without numerical difficulty introduced by the shocks. Vieux and Gaur (1994) presented a distributed watershed model based on this nodal solution using finite elements to represent the drainage network.

A detailed description of the solution methodology used by *r.water.fea* may be found in Vieux (2000). The naming convention stems from the original concept of a geographic information systems (GIS) tool resident within the Geographic Resources

Analysis Support System (GRASS) GIS for simulating surface runoff in watershed.

Research Questions

Research questions related to parameter modeling are the following:

• Parameterizing a distributed hydrologic model requires adequate spatial variability. Improved understanding is needed of the influence on reliability, on calibration, and on prediction biases caused by data characteristics related to resolution, surrogate measures, and temporal resolution of input and parameters.

• New digital elevation datasets are becoming available. Soon, the Shuttle Radar Topography Mission (SRTM) elevation data will provide 30-meter resolution digital elevation models (DEMs) of over 80 percent of the inhabited earth. DEM resolution affects the model responsiveness and flow routing properties. Understanding the influence of resolution and vertical/horizontal precision on hydrologic modeling is needed for efficient model prediction.

• Distributed model calibration is resolution-dependent. Methods are needed for easily moving from one resolution used in model calibration/validation to a greater resolution in the operation of the model.

• Effects of using parameters and inputs at various resolutions on model uncertainty and reliability need investigation, particularly when coarse-resolution rainfall is input to a fine-resolution, distributed-parameter hydrologic model.

Lead-Time and Accuracy of Flood
Forecasts in Small Basins

Radar is a key source of spatially and temporally distributed rainfall data for hydrologic modeling. Vieux (2000) describes the use of weather radar in hydrology with primary emphasis at the river basin scale. Interest in using radar estimates of rainfall in distributed modeling comes from the desire to reduce errors associated with imprecise knowledge of the amount and distribution of rainfall input to models.

Radar and satellite sensors offer improved spatial pattern definition to fill in between rain gauges. However, this source of data has limitations as well. Having access to real-time fusion of rain gauge, satellite, and radar estimates of precipitation would advance hydrologic prediction. Algorithms and systems that can ingest multiple-sensor platforms and produce high-resolution rainfall estimates in space and time are needed to make flood forecasts.

Research Questions:

Research questions related to flood forecasts in small basins include the following:

• How can multisensor rainfall estimates be best fused together given the range of spatial/temporal scales and inherent uncertainties?
• What errors are inherent in each data source, and how do these data characteristics influence flood prediction?
• Considering the rainfall estimation errors over a river basin, what are the considerations necessary to relate space–time scales of the storm in relation to the spatial extent of the basin?
• How do basin characteristics combine and interact with each other to transform rainfall into runoff?

The efficiency of providing warnings in a timely manner must be considered in relation to the time necessary to measure the rainfall, compute the response, and issue the warning. To be of value, there is a limit to the basin size for which forecasts may be made in a timely manner. As the basin becomes smaller, the response time decreases. In order to perform real-time flood forecasting, the rainfall must be estimated by the radar or other sensors, ingested into a hydrologic, model and then compared to some flood stage threshold. Once it is determined that the flood stage is likely to be exceeded, then a warning must be issued to residents and/or emergency personnel. Feasibility of a real-time flood warning system based on radar and model results is highly influenced by the basin size and characteristics coupled with rainfall intensities, duration, and storm track over the basin.

Research on the relationship between storm and basin scale and the time necessary to issue warnings that are reliable and of value is needed so that action can be taken to mitigate flood damages.

Specific Research Questions:

Specific research questions related to flood forecasts in small basins are the following:

• Computational time may be limited when forecasting small basins. Can simplified predictive relationships replace physics-based models in predicting basin response?

• How can a deterministic model be used to forecast flooding when quantitative precipi-tation forecasts are probabilistic?

• What is the limit of predictability in terms of uncertainty in parameters, in model input, and in the resulting model predictions?

• What is the practicable or feasible limit to issuing warnings for small upland basins?

• How does DEM resolution affect the model's predictive capabilities in relationship to various basin sizes ranging from 1 or 2 km^2 to 10, 100, or 1,000 km^2?

Interdependence of Atmospheric, Soil Moisture and Basin Runoff Components

Transforming intense rainfall into runoff depends on the initial state of the soil. The degree to which the basin hydrologic response is affected depends on soil properties, antecedent storm events, and meteorological factors. Some basins may produce about the same amount of runoff whether the soil is wet or dry if clayey or impervious soil conditions prevail. Such basins transform nearly all the rainfall into runoff regardless of initial soil moisture content. Other basins can respond very differently depending on soil moisture. Hydrograph response is also nonlinear to soil moisture contents less than some threshold. At soil moisture contents of greater than 50 percent, a greatly amplified response to soil moisture conditions results.

Research Questions

Research questions related to the interdependence of components include the following:

- Can adjoint methods used in atmospheric models be exploited in physics-based hydrologic models to automate calibration or make real-time adjustments to flood forecasts?
- Do unique values exist for a river basin and a series of storm events?
- What basin characteristics combine to make the adjoint solution ill-conditioned? Can regularization of the cost function improve retrieval of optimal parameters?
- How can we most efficiently couple atmosphere, soil moisture, and basin runoff models to provide flood forecasts for small basins?
- Using global data sets for elevation, soils, and vegetation, can specific basin charac-teristics be identified that will improve model formulations, guide selection of spatial/temporal resolution, and indicate the existence of weak/strong coupling of hydrologic cycle components?

CURRENT STATUS AND OPPORTUNITIES IN QUANTITATIVE PRECIPITATION ESTIMATION AND FORECAST (QPE AND QPF)

W. F. Krajewski
University of Iowa

Methods and systems for QPE

Rain Gauge Based Methods

Research continues on the interpolation methods of spatial patterns but the quantitative differences between different approaches do not seem to be substantial. This is because the error of these approaches is dominated by the spatial sampling error. The geostatistical methods seem to be gaining popularity probably due to their ability to quantify the associated errors. Surprisingly, the Thiessen method (essentially a nearest neighbor method) is still in use although it has been demonstrated many times that its performance is poor.

Much research focused recently on the problem of rain gauge measurement accuracy, in particular the wind effect causing underestimation of rainfall quantities. Both experimental (intercomparison, Legates and DeLiberty 1993; Yang et al. 1998) and numerical (computational fluid dynamics, Nešpor and Sevruk 1999; Habib et al., 1999) approaches are used to investigate the problem. The conclusion is that the underestimation of the long term totals rarely exceed 5 percent for rainfall while snowfall estimates maybe underestimated by as much as 50 percent. However, there are numerous other problems with rain gauge measurements. All gauge designs require frequent maintenance and are subject to mechanical failures while the very popular tipping bucket suffer from their inability to accurately measure lower rainfall amounts (Steiner et al. 2000; Habib et al. 1999). To help in early detection of these problems Ciach and Krajewski (1999a) proposed to use a dual-sensor design. This configuration has been implemented at a number of experimental sites with great success.

Weather Radar Based Methods

The technology has been in use for almost 50 years, yet only recently it found its way into operational use. Currently, the single radar observable used in quantitative estimation of rainfall (operationally) is radar reflectivity. Its conversion to rainfall requires the use of a Z-R relationship. Much of the published research focused on the problem of Z-R selection ignoring numerous and often more significant sources of uncertainty. Relatively little has been reported on the problem of real-time anomalous propagation echo detection (Grecu and Krajewski, 2000c, the effects of the vertical reflectivity profile and their correction, and the effect of the spatial and temporal gradients in the rainfall processes (Kitchen et al., 1994; Andrieu and Creutin, 1995; Joss and Lee, 1995; Anagnostou and Krajewski, 1998; Anagnostou and Krajewski, 1999; Seo et al., 2000).

However, the single most important problem is lack of the rigorous assessment method-ologies. There is no system in place that would systematically monitor progress in the field allowing quantitative comparison of improvements in algorithms in a statistically rigorous and hydrologically relevant way (Ciach and Krajewski, 1999a,b; Ciach et al., 2000). Young et al. (2000) attempted such an evaluation of the hourly product used operationally by the National Weather Service as input to hydrologic models and concluded that they could not find indepen-dent information.

Another important problem is that for proper assessment of radar-rainfall errors we need to know the spatial variability of point rainfall at scales below that of the typical resolution of radar-based products (i.e., 2 km by 2 km). Operational networks have inadequate spacing and past experimental studies (in the 40s, 50s, 60s, and 70s) did not cover such small scales either. Small-scale clusters of rain gauges should be established operationally to collect the relevant information for different rainfall regimes.

The upcoming technology of multiple parameter radar has been demonstrated to be promising but a systematic system for its evaluation is needed otherwise there is a potential that significant investments will be wasted. An experiment Joint Polarization Experiment (JPOLE) planned for 2003 is a major opportunity in this respect.

Extensive use of radar observation in research (e.g., in the context of WEB) has been hampered by difficult and costly access to data. The National Climatological Data Center (NCDC) that archives both the data and the products is not set up for efficient and convenient data distribution. It takes a considerable effort and expertise to assemble a radar-rainfall data set for research on other applications.

Satellite-Based Methods

The research community in the field of rainfall estimation has been very active in the past 10 years (e.g., see Petty and Krajewski, 1996; Petty, 1995). The culmination of their efforts was the Tropical Rainfall Measurement Mission (TRMM) launched in November 1997 (Simpson et al., 1996; Kummerow et al., 1998). The mission continues stimulating much of rainfall research, including research on new rainfall estimation algorithms and their validation. Perhaps the most mature satellite-based rainfall estimation product is the 10 year data set of monthly rainfall produced by the Global Precipitation Climatology Project (GPCP) with resolution of 2.5° by 2.5° (Huffman et al. 1997). An important aspect of the project continues to be the validation of its results. Several algorithm intercomparison projects (AIPs) were conducted and a center for surface reference data has been established to pursue the validation activities (e.g., Arkin and Xie 1994; PIP-1, 1994; Krajewski et al., 2000). The GPCP is being reorganized towards producing a higher resolution global data set (daily accumulations at 1° by 1° spatial resolution).

There are many algorithms documented in the literature and the previous inter-comparison projects do not indicate a clear "winner." To establish the accuracy of the space-based products rigorous framework is needed together with adequate reference data. Depending on the space/time scale of the products the requirements for the type and accuracy of the reference data may vary considerably.

The international research community is planning a successor to TRMM called Global Precipitation Mission (GPM), a constellation of eight orbiting satellites equipped with the same passive suite of sensors calibrated with a space-borne radar. The mission is a major opportunity for studies of global processes and integrated management of water resources.

Methods and Systems for QPF

Classification of the QPF methods is less obvious. For the purpose of this presentation one can distinguish statistical methods and physically-based methods. The focus in the discussion below is on short-term (hours) forecasting methods. The statistical methods are based on empirical data for a variety of observing systems. The methods seek a relation between observed data and the future precipitation quantities. A recent review of the methods based on radar observations was given by Wilson et al. (1998). Also Grecu and Krajewski (2000b) conducted a large-sample study of such methods.

The statistical methods can be (and are) used in operational environment (Zawadzki et al., 1999; Wilson et al., 1998). The main challenge here is to design an experiment that would allow consistent, statistically sound, long-term evaluation of the performance of these methods. Also, as follows from the discussion above on the rainfall estimation methods, it is important to figure out a way to separate the forecasting error that is due to the method from that of the observation that serve as the forecast verification.

The second group of methods is mathematical models of the physical processes involved in generation of precipitation. These models range from detailed cloud physics models to mesoscale models with parameterized cloud physics to simplified vertically integrated models that focus on the precipitation processes. The main challenge here is data assimilation. Major opportunity exists if the difficulty of assimilating radar data can be overcome (e.g., Sun and Crook, 1997; Grecu and Krajewski, 2000a). The approaches currently being explored focus on variational methods. In addition to the issues of computational efficiency, these methods do not explicitly account for the model error. Kalman filter type methods provide a framework in which both the modeling error and the observational error can be acknowledged and quantita-tively accounted for. However, at this time feasibility of such methods is questionable and the potential benefits difficult to assess. Filtering methods can be used (Georgakakos, 2000) with a simpler class of physically-based models such as those proposed by Lee and Georgakakos (1996) and French and Krajewski (1994). Again, well design comparative experiments are needed to properly assess the performance of the complex and simple models.

Recommendations: QPE

• Deploy dual-sensor rain gauge designs in the operational environment.

• Establish clusters of rain gauges for observations of small-scale rainfall variability.

• Develop benchmarks of radar-rainfall estimation performance for monitoring progress in algorithm development and assessment of new technologies.

• Develop methodologies of using rain gauge data based reference standards for evalua-tion of radar-rainfall estimates.

• Develop new sensors and technologies that would provide a better reference standard than the gauges.

• Improve access to radar-rainfall data for the research community.

Recommendations: QPF

- Develop a set of benchmarks based on large samples of data.
- Develop a framework for evaluating the quality of forecasts independent of the errors of the reference data.
- Conduct comparative studies of various models and methods in a consistent way.

ISSUES OF SCALE ON PRECIPITATION PREDICTABILITY

E. Foufoula-Georgiou
University of Minnesota

INTRODUCTION

It is well known that some factors affecting predictability of atmospheric and hydrologic variables are related to: uncertainty in the initial conditions, uncertainty in the boundary conditions, and inadequate knowledge of the system dynamics and multiscale interactions of the involved variables. Efforts aimed at quantitatively determining the range of predictability in the atmospheric-hydrologic system have been based either on idealized systems of dynamic equations or in model predictability studies employing more elaborate numerical models. In the first case, methods of no-linear dynamics (e.g., Lyapunov exponents) or methods of stochastic propagation of uncertainty have been used to assess predictability. In the second case, predictability is assessed by quantitative comparison of model forecasts with observations (assuming that the model is completely known and true). In both cases, several issues arise that are far from resolved and need a focused research program to systematically address them.

In this talk, the following questions will be tackled both conceptually and via examples. Results are mostly restricted to those obtained in our group and thus this note is far from comprehensive even in the questions that it tries to address.

Question 1: *How do the feedback characteristics of two-way coupled land-atmosphere models impact predictability?*

Answer: *The feedback characteristics amplify uncertainties (due to initial conditions, boundary conditions, and incomplete knowledge of small-scale structure) and thus limit predictability.*

Supporting Evidence:

It has been found that the nonlinear feedback dynamics of the coupled land-atmosphere system amplify the small-scale (1-5km) variability of the forcing variables to create larger-scale variability on other predicted water and energy fluxes (Nykanen et al., 2000). Such an effect imposes limits on the predictive power of numerical hydrometeorological models even if the interest is at larger scales (see also the discussion in Lorenz, 1969). These results suggest that the small-scale variability of precipitation and its nonlinear propagation through the land-atmosphere system cannot be ignored if accurate predictions are desired. Variability of precipitation at scales less than 5km can be incorporated in coupled models in three ways: (a) explicitly running the models at resolution of less than 5 km over the whole domain of interest even if larger scale predictions are requiredd; (b) run nested models which have lower resolution at the outer domain and less than 5 km resolution in the inner domain which represents the area of interest; and (c) run the models at a resolution that can account for the mesoscale atmospheric dynamics (20-30 km) and statistically enhance the variability and resolution of precipitation while preserving large-scale averages and explicitly propagating this enhanced variability (via fine scale two-way feedbacks) through the land-atmosphere system. Option (a) is computa-tionally infeasible in most of today's computers and for most researchers and is certainly prohibiting in operational forecasts. Option (b) has its own problems, which remain mostly unresearched to date. A main problem is the propagation of boundary conditions (e.g., see Chu, 1999), which manifests itself by giving different predictions depending on the resolution of the outer domain, the convective parameterization of the outer domain, and the size and position of the inner nested domain. This forms one open problem in which research has to be accelerated.

It has been found that the small-scale variability of cloud hydrometeors and precipitation (down to less than 1 km scale) considerably affects the radiative trasfer through simulated clouds (see Harris and Foufoula-Georgiou, 2000) and thus precipitation retrievals from microwave sensors. Typical algorithms for microwave retrieval of precipitation are based on inversion of a data base obtained from cloud model simulations (typically run at 3 km resolution) and simulated brightness temperature computed via radiative transfer. It has been found that cloud models run at 3 km resolution miss the proper variability at scales up to five times the model resolution (see Harris et al, 2000). Not accounting for this variability affects the computed brightness temperature and the estimated precipitation at large scales (10 to 50 km estimates depending on the channel frequency). This in turn has an impact on climate studies which use these tropical rain estimates for data assimilation or climate model verification.

Question 2: *How can the predictive power of models (which remains one practical tool for assessing system predictability) be assessed based on observations?*

Answer: *QPF verification methodologies, currently in use, present problems and new methodologies that account for scale issues must be developed for an accurate assessment of the predictive power of models.*

Supporting Evidence:

Typical measures of performance, such as threat score (TS) and root mean square error (RMSE) have been shown to be of limited power to depict important differences between the forecasted and observed precipitation. A suite of multiscale methods for comparison of forecasted precipitation to observed precipitation has been developed and has shed some light into shortcomings of numerical weather prediction models in terms of capturing statistical properties of precipitation which are considered important in the prediction of other fluxes in the land-atmosphere system (Zepeda-Arce et al., 2000). Some new measures of forecast verification which are more appropriate for highly variable fields have been explored, such as the Hausdorff distance between two images and have been found to have potential for further study (Dodov and Foufoula-Georgiou, unpublished manuscript, 2000).

A popular method of QPF verification based on RMSE between the outputs of the model and point observations by transforming the areal averages to point values (area to point) or the point values to areal averages (point to area) has been critically assessed and has been found to suffer from scale problems. In particular, changing the scale of the observations to match the scale of the model output (point to area) or vice versa (area to point) imposes a "representa-tiveness error" which is nonzero even in the case of a perfect model (i.e., a model whose predictions are indeed the true averages of the underlying field) and moreover is dependent on scale. It has been shown that the magnitude of the representativeness error is significant (it can be up to 50 percent of the spatial average of the precipitation filed) and has considerable scale dependency within the typical mesoscale range of 5 to 50 km (Tustison et al, 2000). In verification studies in which model performance is assessed as a function of model resolution, ignoring or mischaracterizing the scale dependent representativeness error can significantly affect inferences about the model performance as a function of scale. This is because the total error is composed on the observational error, the representativeness error, and the model error, and thus misspecifying the representativeness error directly mispecifies the model error and any inferences about the model performance as a function of resolution and thus predictability inferences in general. Some new methodologies are under exploration, which are based on multiscale filtering and can explicitly account for scale effects.

Question 3: *Theoretical studies of predictability and uncertainty propagation across scales require that semi-empirical parameteri-*

zations are appropriately scaled such that fluxes are preserved. How is this to be accomplished?

Answer: *Commonly used relationships must be studied to establish "scaling" relationships, which depend on the form of the nonlinear relationship and the spatial variability of the involved variables.*

Supporting Evidence:

Through the analysis of a commonly used relationship that parameterizes surface runoff as a function of soil moisture, it has been demonstrated how the parameters of the relationship must change with scale such that spatially averaged predicted fluxes can be preserved at any scale of interest (Nykanen and Foufoula-Georgiou, 2000). Similar analyses can be performed for other relationships such that the parameters are directly related to the scale-dependent variance of the underlying field and the form of the parameterization. In our analysis, data from the 1997 Southern Great Plains Hydrology Experiment (SGP97) have been used to quantify the spatial variability of soil moisture and it was shown that changing the scale (resolution) at which the model is applied within reasonable ranges (5 km to 50 km) while keeping the parameterization the same can result in biases up to 30 percent in predicted runoff. These issues are not only of practical importance for more accurate predictions but of theoretical importance too in studies that will attempt addressing predictability questions based on the dynamical equations of the land-atmosphere system.

WHY AND HOW DOES PRECIPITATION CHANGE?

K. E. Trenberth
National Center for Atmospheric Research

INTRODUCTION

Why does it rain? If a parcel of air rises, it expands in the lower pressure, cools, and therefore condenses moisture in the parcel, producing rainfall—or perhaps snowfall. So key ingredients are certainly the many and varied mechanisms for causing air to rise. These range from orographically uplifted air as air flows over mountain ranges to a host of instabilities in the atmosphere that arise from unequal heating of the atmosphere, to potential vorticity dynamics. The instabilities include those that result directly in vertical mixing such as convective instabil-ities, to those associated with the meridional heating disparities that give rise to baroclinic instabilities and the ubiquitous weather systems. Thus cold air pushing underneath warmer air (advancing cold front) or warm air gliding over colder air (advancing warm front), and so on, can all provide opportunities for air to rise. These mechanisms for causing air to rise presumably do not in of themselves change. They need to be better understood and modeled, but from a societal standpoint the fact that their relative importance can change with time is of more consequence. These aspects have been the focus of a lot of meteorological research and are only briefly addressed here.

The other main ingredient in the opening statement about why it rains is the assumption that there is moisture present.

Where exactly does the moisture come from? It is argued here that this aspect of precipitation is one that has been under appreciated and is worthy of more attention. After all it will not rain at all unless there is a supply of moisture.

Questions and Issues

1. Better documentation and processing of all aspects of precipitation. The needs include frequency, intensity, and amount. Trenberth (1998) has argued for the creation of a data base of these quantities using hourly precipitation amounts. This time interval averages over individual cells within a storm but typically allows the evolution of a storm to be grossly captured. It is compatible with the time steps in global models, which are typically two or three per hour. It is viable from a data management standpoint. It is also viable from many observations, from recording rain gauges and from radar (e.g., NEXRAD) and satellite (e.g., TRMM, GOES) estimates. In fact, remote sensing measures the instantaneous rain rate, not amount over time, and this has to be converted into a daily amount (e.g., by fitting of log-normal distributions; Short et al., 1993a,b, also Shimizu et al., 1993). Instead it would be better converted into an hourly rate, and histograms of the rate then provide the basic information database.

2. Increased understanding of the efficiency of precipitation and how it changes with environmental conditions. "Precipitation efficiency'" is defined as the ratio of the water mass precipitated to the mass of water vapor entering the storm through its base (e.g., Fankhauser 1988) or the ratio of total rainfall to total condensation in modeling studies (e.g., Ferrier et al. 1996). These studies show that in typical storms the precipitation efficiency varies from about 20 to 50 percent, and 30 percent seems a typical value. In one cloud/thunderstorm model, greater moisture content in the atmosphere produced more condensation and greater precipitation efficiency (Ferrier et al., 1996). However, warmer conditions, which might accompany greater moister content could also imply that more moisture might remain if relative humidity is a key factor, as is likely. Therefore the rainfall may not increase in direct

proportion to the moisture convergence because more moisture is left behind. The efficiency of precipitation in climate models is implicit and has not been addressed. Pollution and aerosols may influence the microphysics of precipitation processes and also influence the efficiency of precipitation.

3. Improved parameterization of convection in large-scale models. Parameterization of convection needs to be improved to appropriately allow Convective Available Potential Energy (CAPE) build up as observed and may involve "triggers" and thresholds to initiate convection. This seems to be a scale interaction problem in part, as larger scale motions may suppress or enhance the release of CAPE. As well as the diurnal cycle it may also be a key in the Madden-Julian Oscillation in the tropics which has been difficult for models to get right.

4. Improved observations and modeling of sources and sinks of moisture for the atmosphere, especially over land. This relates to recycling and the disposition of moisture at the surface in models, and whether the moisture is or is not available for subsequent evapotranspira-tion. It relates to improved and validated treatment of runoff, soil infiltration and surface hydrology in models as well as vegetation models.

5. Improved analysis of precipitation rates. We have argued (Trenberth, 1998, 1999a) that increasing the moisture content of the atmosphere should increase the rate of precipitation locally by invigorating the storm through latent heat release and further by supplying more moisture, although what happens to the amount is less clear, as the duration of a storm may also be shortened. Some analysis of model results supports this view but most analyses have used daily and not hourly or higher frequency data, so they also highlight the need for more attention to the nature of the analysis of both models and observational datasets.

6. Improved analysis of the frequency of precipitation and changes in weather systems. Changes occur in large-scale atmospheric circulation, extratropical storms and the overall baroclinicity, from year to year and especially as the climate changes. Held (1993) notes that extratropical storms are greatly influenced by moisture in the atmosphere and that one effect of increased moisture content in the atmosphere is to enhance the latent heating in

such storms and thereby increase their intensity. On the other hand, he also notes that more moist air would be transported pole-wards by transient eddies, reducing the required poleward energy transports normally accomplished by baroclinically unstable eddies and the associated poleward down-gradient heat transports. He therefore argues that this would contribute to "smaller eddies" and suggests that this means a decrease in eddy amplitudes. While recognizing that both effects are important, Held suspects that the latter is dominant. There are other possibilities not considered by Held. In particular, individual storms could be more intense from the latent heat enhancement, but fewer and farther between (Trenberth, 1998), and so this relates to the frequency of storms and precipitation. Changes in the vertical temperature structure (the lapse rate) will also play a role in such storms. Therefore another major factor worth considering in more detail is the frequency and nature of precipitation events. CLIVAR is designed to address other aspects of changes and variability in atmospheric circulation.

7. Improved simulation of the diurnal cycle of precipitation in models. This probably also requires improved simulation of the diurnal cycle of temperature, cloud amount and atmospheric circulation, as well, and especially the build up and release of CAPE.

8. Improved ocean modeling. Rainfall amounts over the oceans are also important because of its contribution to the fresh water budget for the ocean—or equivalently salinity. It is a key in the potential changes in the thermohaline circulation as the climate changes (Gent 2000). Part of this also relates to runoff from land, but it is evaporation and precipitation over the oceans, that is most important along with mass transports and mixing within the ocean.

9. Moisture sources for storms. The above back-of-the-envelope calculations on how far afield the storm-scale circulation reaches out to gather in the moisture that fuels the storm and provides the precipitation probably vary greatly with synoptic situation, with season, and with the nature of the storm and mechanism for producing the precipitation. Can, these numbers be consolidated and refined, and it is worthwhile?

EVOLUTION OF REGIONAL HYDROLOGIC PROCESSES WITHIN THE CLIMATE SYSTEM AS AN INITIAL VALUE PROBLEM: AN OVERVIEW

R. A. Pielke Sr.
Colorado State University

The "hydrologic system" is an intimately coupled component of the Earth's climate system. It is more appropriate, therefore, to discuss hydrologic processes as a component of the climate, rather than as "a hydrologic system". Interactions within the climate system occur at a wide range of temporal and spatial scales, as discussed, for example, in Pielke (1998). In the context of prediction, the interactions that involve water with other components of the climate results in an initial value problem, if the feedbacks are sufficiently large and nonlinear, and occur within the time period which the forecasts are made for. A consequence of these nonlinear interactions is that the climate system is inherently chaotic. In my talk, the concept of climate as an initial value problem will be illustrated with respect to land-atmosphere feedbacks, first with simple models (Zeng et al., 1990; Pielke and Zeng, 1994), and then with several more physically realistic coupled regional and global models of the Earth system (Chase et al., 1999; Pielke et al., 1999; Eastman et al., 2000; Lu et al., 2000). In the context of land-atmosphere interactions, these feedbacks range from short term biophysical, to medium range biogeochemical, to long-term biogeographical interactions. Hydrologic processes are involved on each of these time scales. These conclusions are being confirmed by other investigators as reported, for example, in Pitman et al. (1999), Chen et al. (2000), and the numerous reference citations in Pielke (2000). Among the major conclusions thus far are:

1. Hydrologic processes must be coupled with carbon and nitrogen processes (and those of other trace gases and aerosols) in order to develop a realistic understanding of the climate. The concept of WEB (Water-Earth-Biota; Gupta et al. 2000:http://cires.-colorado.edu/hydrology), which has been initiated by the Hydrology Division of the NSF is an initiative to integrate these processes within a research framework.

As an example of this need, with respect to the concept of carbon sequestration (which is designed to reduce the net input of the radiatively active gas carbon dioxide into the atmosphere), there has been no assessment of whether water vapor, another radiatively active gas, would have increased input into the air as a result of this carbon mitigation approach. The net result of a carbon sequestration program that does not also consider the associated effect on hydrologic processes could be a greater net flux of radiatively active gases into the atmosphere (as well as a change of the net radiation at the surface by altering its albedo) than would occur without carbon sequestration!

2. The biological effect of CO_2 on ecosystem function and feedback into atmospheric dynamics and thermodynamics must be included in the assessment of climate and weather prediction. In the short time period, for example, locally increased carbon dioxide associated with urban pollution can alter the stomatal resistance to transpiration loss. This reduction in water flux can influence convective available potential energy (CAPE), and thus subsequent rainfall. On long time scales, the transient evolution to increased carbon dioxide can affect vegetation species composition (more C3 plants at the expense of C4 plants, for example), which subsequently feedbacks to alter rainfall amounts and patterns, as shown by Eastman et al. (2000).

3. Human caused landscape disturbance is a major perturber of the Earth's climate system on the global scale. Indeed, it appears to already have had an effect on the Earth's climate system

that is as large or larger than claimed for the radiative effect of doubled CO2. Landscape change has been accelerating and our largest alterations have occurred in recent years (Leemans 1999; O'Brien 2000). This conclusion requires that we must include landscape as an initial value within models of the climate.

4. The success of seasonal weather forecasts has primarily been a result of the treatment of the sea surface temperatures as a static lower boundary condition (Landsea and Knaff 2000). These seasonal "nowcasts" work because the dominant atmospheric-ocean feedbacks occur over a relatively long time (i.e., longer than a season). Similarly, antecedent soil moisture provides a long enough memory for useful seasonal nowcasts (such an inertia within the climate system can help explain the persistent of the Texas drought and heat this summer, since the vegetation in the region is not transpiring due to poor soil moisture). The lack of skilled multi-year SST predictions apparently results because the forecasts on this time period are no longer nowcasts, but true initial value problems with the inherent limitation on the ability to fore-cast the future.

For these reasons, the concept of weather prediction as an "initial value problem" while "climate is a boundary problem" is being replaced with the new paradigm that weather occurs over a short enough time period so that many of the feedbacks within the climate system do not occur. Both weather and climate are, in fact, initial value problems. "Weather" should be viewed as one subset of the "climate" system.

REFERENCES FOR PRESENTED PAPERS

Anagnostou, E. N., and W. F. Krajewski. 1998. Calibration of the NEXRAD precipitation processing subsystem. Weather and Forecasting 13:396-406.

Anagnostou, E. N., and W. F. Krajewski. 1999. Real-time radar rainfall estimation. Part 1: algorithm formulation. Journal of Atmospheric and Oceanic Technology 16:189-197.

Andrieu H., and J. D. Creutin. 1995. Identification of vertical profiles of reflectivities for hydrological applications using an inverse method. Part I: formulation. J. Appl. Meteor. 34:225-239.

Arkin, P. A., and P. Xie. 1994. The global precipitation climatology project: First algorithm intercomparison project. Bull. Am. Meteorol. Soc. 75:401-419.

Chen, F., R. A. Pielke, Sr., and K. Mitchell. 2000. Development and application of land-surface models for mesoscale atmospheric models: Problems and promises. AMS Monograph.

Chu, P. 1999. Two kinds of predictability in the Lorenz system. J. Atmos. Sciences 56:1427-1432.

Ciach, J. G., and W. F. Krajewski. 1999a. On the estimation of radar rainfall error variance. Advances in Water Resources 22:585-595.

Ciach, G. J., and W. F. Krajewski. 1999b. Conceptualization of radar-raingage comparisons under observational uncertainties. Journal of Applied Meteorology 38:519-1525.

Ciach, G. J., M. L. Morrissey, and W. F. Krajewski. 2000. Conditional bias in radar rainfall estimation. Journal of Applied Meteorology.

Crook, N. A. 1996. Sensitivity of moist convection forced by boundary layer processes to low-level thermodynamic fields. Mon. Wea. Rev. 124:1767-1785.

Delworth, T., and S. Manabe. 1988. The influence of potential evaporation on the variabilities of simulated soil wetness and climate. J. Climate 1:523-547.

Eastman, J. L., M. B. Coughenour, and R. A. Pielke. 2000. The effects of CO_2 and landscape change using a coupled plant and meteorological model. Global Change Biology.

Epstein, E. S. 1969. Stochastic dynamic prediction. Tellus 21:739-759.

Errico, R., and D. Stensrud. 2000. Estimation of Error Statistics of Precipitation Produced by Convective Parameterization Schemes for Application to the Variational Assimilation of Precipitation Observations. Draft Manuscript.

Fankhauser, J. C. 1998. Estimates of thunderstorm precipitation efficiency from field measurements in CCOPE. Monthly Weath. Rev. 116:663-684.

Ferrier, B. S., J. Simpson, and W-K Tao. 1996. Factors responsible for precipitation efficiencies in midlatitude and tropical squall simulations. Monthly Weath. Rev. 124:2100-2125.

French, M. N., and W. F. Krajewski. 1994. A model for real-time quantitative rainfall forecasting using remote-sensing, 1, formulation. Water Resources Research 30:1075-1083.

Gent, P. R. 2000. Will the North Atlantic Ocean Thermohaline Circulation Weaken During the Next Century?

Georgakakos, K. P. 2000. Covariance propagation and updating in the context of real-time radar data assimilation by quantitative precipitation forecast models. Journal of Hydrology.

Grecu, M., and W. F. Krajewski. 2000a. Rainfall forecasting using variational assimilation of radar data in numerical cloud models. Advances in Water Resources.

Grecu, M., and W. F. Krajewski. 2000b. A Comprehensive investigation of statistical procedures for radar-based short-term, quantitative precipitation forecasting, Journal of Hydrology.

Grecu, M., and W. F. Krajewski. 2000c. An efficient methodology for detection of anomalous propagation echoes in radar reflectivity data using neural networks. Journal of Oceanic and Atmospheric Technology 17:121-129.

Habib, E., W. F. Krajewski, V. Nešpor, and A. Kruger. 1999. Numerical simulation studies of raingage data correction due to wind effect. Journal of Geophysical Research-Atmospheres Research 104:(19)723-734.

Harris, D., and E. Foufoula-Georgiou. 2000. Subgrid variability and stochastic downscaling of modeled precipitation and its effects on radiative transfer computations. J. Geophys. Res.

Harris, D., E. Foufoula-Georgiou, K. Droegemeir and J. Levit. 2000. Multiscale statistical properties of a high resolution precipitation forecast. J. Hydrometeorology.

Held, I. M. 1993. Large-scale dynamics and global warming. BAMS 74:228-241.

Huang, J., and H. M. van den Dool. 1993. Monthly precipitation-temperature relation and temperature prediction over the U.S. J. Climate 6:1111-1132.

Huang, J., H. M. van den Dool, and K. G. Georgakakos. 1996. Analysis of model-calculated soil moisture over the U.S. (1931-1993) and applications to long range temperature forecasts. J Climate. 9:1350-1362.

Huffman, G. J., R. F. Adler, P. Arkin, A. Chang, R. Ferraro, A. Gruber, J. Janowiak, A. McNab, B. Rudolf, and U. Schneider. 1997. The global precipitation climatology product (GPCP) combined precipitation data set. Bull. Am. Meteorol. Soc. 78:5-20.

Joss, J., and R. Lee. 1995. The application of radar-gauge comparisons to operational profile corrections. J. Appl. Meteor. 34:2612-2630.

Kitchen M., R. Brown, A. G. Davis. 1994. Real time correction of weather radar for effects of bright band, range and orographic growth in widespread precipitation. Q.J.R. Meteorol. Soc. 120:1231-1254.

Koster, R. D., and P. C. D. Milly. 1997. The interplay between transpiration and runoff formulations in land surface schemes used with atmospheric models. J . Climate. 10:1578-1591.

Koster, R. D., M. J. Suarez, and M. Heiser. 2000. Variance and predictability of precipitation at seasonal-to-interannual timescales. J. Hydrometeor. 26-46.

Krajewski, W. F., G. J. Ciach, J. R. McCollum, and C. Bacotiu. 2000. Initial validation of the global precipitation climatology project over the United States. Journal of Applied Meteorology 39:1071-1086.

Kumar, A., and M. P. Hoerling. 1995. Prospects and limitations of seasonal atmospheric GCM predictions. Bull. Amer. Met. Soc. 335-345.

Kummerow, C., W. Barnes, T. Kozu, J. Shiue, and J. Simpson. 1998. The tropical rainfall measuring mission (TRMM) sensor package. Journal of Atmospheric and Oceanic Technology 15:809–817.

Landsea, C. W., and J. A. Knaff. 2000. How much skill was there in forecasting the very strong 1997-98 El Nino? Bull. Amer. Meteor. Soc. 81:2107-2119.

Lee, T.-H., and K. P. Georgakakos. 1996. Operational rainfall prediction on meso-gamma scales for hydrologic applications. Water Resources Research 32:987-1003.

Leemans, R. 1999. Land-use change and the terrestrial carbon cycle. IGBP Global Change Newsletter 37:24-26.

Legates, D. R., and T. L. DeLiberty. 1993. Precipitation measurement biases in the United States. Water Resources Bulletin 29:855-861.

Leith, C. E. 1974. Theoretical skill of Monte-Carlo forecasts. Mon. Wea. Rev. 102:409-418.

Liu, Y., and R. Avissar. 1999. A study of persistence in the land-atmosphere system using a general circulation model and observations. J. Clim. 2139-2153.

Lohmann, D., and co-authors. 1998. The project for the intercomparison of land-surface parameterization schemes (PILPS) phase 2c Red Arkansas river basin experiment: 3. spatial and temporal analysis of water fluxes. Glob. and Planet. Change 19:161-180.

Lorenz, E. N. 1975. Climate predictability. The physical basis of climate modeling. WMO GARP Publication Series 16:132-136.

Lorenz, E. 1969. The predictability of a flow, which possesses many scales of motion. Tellus 3:289-307.

Lovelock, J. 1995. The Ages of Gaia. Oxford University Press, Oxford, UK.

Lu, L., R. A. Pielke, G. E. Liston, W. J. Parton, D. Ojima, and M. Hartman. 2000. Imple-mentation of a two-way interactive atmospheric and ecological model and its application to the central United States. J. Climate.

Nešpor, V., and B. Sevruk. 1999. Estimation of wind-induced error of rainfall gauge measurements using a numerical simulation. Journal of Atmospheric and Oceanic Technology 16:450-464.

Nykanen D, E. Foufoula-Georgiou, and W. Lapenta. 2000. Impact of small-scale precipitation variability on larger-scale spatial organization of land-atmosphere fluxes. J. of Hydro-meteorology.

Nykanen D., and E. Foufoula-Georgiou. 2000. Soil moisture variability and its effect on scale-dependency of nonlinear parameterizations on coupled land-atmosphere models. Advances in Water Resources.

O'Brien, K. L. 2000. Upscaling tropical deforestation: Implications for climate change. Climatic Change 44:311-329.

Petty, G., and W. F. Krajewski. 1996. Satellite estimation of precipitation. Hydrological Sciences Journal 41:433-451.

Petty, G. W. 1995. The status of satellite-based rainfall estimation over land. Remote Sens. Environ. 51:125-137.

Pielke, R. A., and X. Zeng. 1994. Long-term variability of climate. J. Atmos. Sci. 51:155-159.

Pielke, R. A. 1998. Climate prediction as an initial value problem. Bull. Amer. Meteor. Soc. 79:2743-2746.

Pielke, R. A. 2000. Influence of the spatial distribution of vegetation and soils on the prediction of cumulus convective rainfall. Rev. Geophys.

Pielke, R. A., G. E. Liston, J. L. Eastman, L. Lu, and M. Coughenour. 1999. Seasonal weather prediction as an initial value problem. J. Geophys. Res. 104:19463-19479.

Pielke, R., Sr., K. Reckhow, and F. Swanson. 2000. New Interdisciplinary Initiative Combines Water, Earth, and Biota, Trans. AGU (EOS).

PIP-1. 1994. The first WetNet precipitation intercomparison project. Remote Sensing Reviews 11:373.

Pitman, A., R. Pielke Sr., R. Avissar, M. Claussen, J. Gash, and H. Dolman, The role of the land surface in weather and climate: Does the land surface matter. IGBP Newsletter, 39, 4-11, 1999.

Reiners, W. A. 1988. Complementary models for ecosystems. American Naturalist 127:59-73.

Ross, R. J., and W. P. Elliot. 1996. Tropospheric water vapor climatology and trends over North America: 1973-1993. J. Clim. 9:3561-3574.

Schlosser, C. A., and P. C. D. Milly. 2000. The potential impact of soil moisture initialization on soil moisture predictability and associated climate predictability. Proceedings of the GEWEX/BAHC International Workshop on Soil Moisture Monitoring, Analysis, and Prediction for Hydrometeorological and Hydroclimatological Applications. Pg. 31.

Seo, D. J., J. P. Breidenbach, R. Fulton, D. Miller and T. O'Banon. 2000. Real time adjustment of range dependent biases in WSR-88D rainfall estimates due to nonuniform vertical profile of reflectivity. Journal of Hydrometeorology.

Shimizu, K., D. A. Short, and B. Kedem. 1993. Single- and double-threshold methods for estimating the variance of area rain rate. J. Meteor. Soc. Japan 71:673-683.

Short, D. A., K. Shimizu, and B. Kedem. 1993a. Optimal thresholds for the estimation of area rain rate moments by the threshold method. J. Appl. Meteor. 32:182-192.

Short, D. A., D. B. Wolff, D. Rosenfeld, and D. Atlas. 1993b. A study of the threshold method utilizing rain gauge data. J. Appl. Meteor. 32:1379-1387.

Shukla, J. 1998. Predictability in the midst of chaos: A scientific basis for climate forecasting. Science 728-731.

Simpson, J., C. Kummerow, W.-K. Tao, and R. F. Adler. 1996. On the tropical rainfall measuring mission (TRMM). Meteor. Atmos. Phys. 60:19-36.

Steiner, M., J. A. Smith, S. J. Burges, C.V. Alonso, and R. W. Darden. 1999. Effect of bias adjustment and rain gauge data quality control on radar rainfall estimation. Water Resour. Res. 35(8):2487-2503.

Sun, J.-Z., and Crook, N. A. 1997. Dynamical and microphysical retrieval from Doppler radar observations using a cloud model and its adjoint. 1. Model development and simulated data experiments. Journal of the Atmospheric Sciences 54:1642-1661.

Trenberth, K. E. 1998. Atmospheric moisture residence times and cycling: Implications for rainfall rates with climate change. Climatic Change 39:667-694.

Tustison, B., D. Harris, and E. Foufoula-Georgiou. 2000. Scale issues in verification of precipitation forecasts. J. Geophys. Res.

Vieux, B. E., V. F. Bralts, L. J. Segerlind, and R. B. Wallace. 1990. Finite element watershed modeling: One-dimensional elements. J. of Water Resources Management Planning 116:803-819.

Vieux, B. E. 1991. Geographic information systems and nonpoint source water quality modeling. J. of Hydrological Processes 5:110-123.

Vieux, B. E. 1993. DEM aggregation and smoothing effects on surface runoff modeling. J. of Computing in Civil Engineering 7:310-338.

Vieux, B. E., and N. Gaur. 1994. Finite element modeling of storm water runoff using GRASS GIS. Microcomputers in Civil Engineering 9:263-270.

Vieux, B. E., F. LeDimet, and D. Armand. 1998. Optimal control and adjoint methods applied to distributed hydrologic model calibration. Proceedings of Int. Assoc. for Computational Mechanics, IV World Congress on Computational Mechanics, 29 June-2 July, Buenos Aires, Argentina. Pg. II1050.

Vieux, B. E. 2000. Distributed Hydrologic Modeling Using GIS. Kluwer Academic Publishers, Spuilboulevard 50, Dordrecht, The Netherlands. Water Science and Technology Series.

Vinnikov, K. Y., A. Robock, N. A. Speranskaya, and A. Schlosser. 1996. Scales of temporal and spatial variability of midlatitude soil moisture. J. Geophys. Res. 7163-7174.

Webster, P. J. 1994. The role of hydrological processes in ocean-atmosphere interactions. Reviews of Geophysics 32:427-436.

Wilson, J. W., N. A. Crook, C. K. Mueller, J. Sun, and M. Dixon. 1998. Nowcasting thunderstorms: A status report. Bulletin of the American Meteorological Society 79:2079-2099.

Yang, D., B. E. Goodisson, and J. R. Metcalfe. 1998. Accuracy of NWS 8" standard nonrecording precipitation gauge: Results and application of WMO intercomparison. Journal of Atmospheric and Oceanic Technology 15:54-67.

Young, B., A. A. Bradley, W. F. Krajewski, and A. Kruger. 2000. An evaluation study of NEXRAD multisensor precipitation estimates for operational hydrologic forecasting. Journal of Hydrometeorology 1:241-254.

Zawadzki, I., J. Morneau, and R. Laprise. 1999. Predictability of precipitation patterns--An operational approach. Journal of Applied Meteorology 33:1562-1571.

Zeng, X., R. A. Pielke, and R. Eykholt. 1990. Chaos in Daisyworld. Tellus 42B:309-318.

Zepeda-Arce, J., E. Foufoula-Georgiou, and K. Droegemeier. 2000. Space-time rainfall organization and its role in validating quantitative precipitation forecasts, J. Geophys. Res. 105(10):129-146.

Zhai, P., and R. E. Eskridge. 1997. Atmospheric water vapor over China. J. Clim. 10:2643-2652.

Appendix C
National Research Council Board
Membership and Staff

WATER SCIENCE AND TECHNOLOGY BOARD

RICHARD G. LUTHY, *Chair*, Stanford University, California
JOAN B. ROSE, *Vice Chair*, University of South Florida, St. Petersburg
RICHELLE M. ALLEN-KING, Stanford University, California
GREGORY B. BAECHER, University of Maryland, College Park
KENNETH R. BRADBURY, Wisconsin Geological and Natural History Survey, Madison
JAMES CROOK, CH2M Hill, Boston, Massachusetts
EFI FOUFOULA-GEORGIOU, University of Minnesota, Minneapolis
PETER GLEICK, Pacific Institute, Oakland, California
STEVEN P. GLOSS, University of Wyoming, Laramie
JOHN LETEY, JR., University of California, Riverside
DIANE M. MCKNIGHT, University of Colorado, Boulder
CHRISTINE L. MOE, Emory University, Atlanta, Georgia
RUTHERFORD H. PLATT, University of Massachusetts, Amherst
JERALD L. SCHNOOR, University of Iowa, Iowa City
LEONARD SHABMAN, Virginia Polytechnic Institute and State University, Blacksburg
R. RHODES TRUSSELL, Montgomery Watson, Pasadena, California

Staff

STEPHEN D. PARKER, Director
LAURA J. EHLERS, Senior Staff Officer
JEFFREY W. JACOBS, Senior Staff Officer
WILLIAM S. LOGAN, Senior Staff Officer
MARK C. GIBSON, Staff Officer
M. JEANNE AQUILINO, Administrative Associate
ELLEN A. DE GUZMAN, Research Associate
PATRICIA JONES KERSHAW, Study/Research Associate
ANITA A. HALL, Administrative Assistant
ANIKE L. JOHNSON, Project Assistant
JON Q. SANDERS, Project Assistant

BOARD ON ATMOSPHERIC SCIENCES AND CLIMATE

ERIC J. BARRON, *Chair,* Pennsylvania State University, University Park
SUSAN K. AVERY, University of Colorado, Boulder
RAYMOND J. BAN, The Weather Channel, Inc., Atlanta, Georgia
HOWARD B. BLUESTEIN, University of Oklahoma, Norman
STEVEN F. CLIFFORD, National Oceanic and Atmospheric Administration, Boulder, Colorado
GEORGE L. FREDERICK, Radian Electronic Systems, Austin, Texas
JUDITH L. LEAN, Naval Research Laboratory, Washington, D.C.
MARGARET A. LEMONE, National Center for Atmospheric Research, Boulder, Colorado
MARIO J. MOLINA, Massachusetts Institute of Technology, Cambridge
ROGER A. PIELKE, JR., National Center for Atmospheric Research, Boulder, Colorado
MICHAEL J. PRATHER, University of California, Irvine
WILLIAM J. RANDEL, National Center for Atmospheric Research, Boulder, Colorado
ROBERT T. RYAN, WRC-TV, Washington, D.C.
THOMAS F. TASCIONE, Sterling Software, Inc., Bellevue, Nebraska
ROBERT A. WELLER, Woods Hole Oceanographic Institution, Massachusetts
ERIC F. WOOD, Princeton University, New Jersey

Staff

ELBERT W. (JOE) FRIDAY, JR., Director
LAURIE S. GELLER, Program Officer
PETER A. SCHULTZ, Senior Program Officer
VAUGHAN C. TUREKIAN, Program Officer
DIANE GUSTAFSON, Administrative Assistant
ROBIN MORRIS, Financial Associate

Appendix D

Biographical Sketches of Members of Committee on Hydrologic Science

Dara Entekhabi *(chair)* is an associate professor in the Department of Civil and Environ-mental Engineering and the Department of Earth, Atmospheric and Planetary Sciences at the Massachusetts Institute of Technology. His research interests are in the basic understanding of coupled surface, subsurface, and atmospheric hydrologic systems that may form the bases for enhanced hydrologic predictability. Specifically, he conducts research in land–atmosphere interactions, remote sensing, physical hydrology, operational hydrology, hydrometeorology, groundwater–surface water interaction, and hillslope hydrology. He received his B.A. and M.A. degrees from Clark University. Dr. Entekhabi received his Ph.D. degree in civil engineering from the Massachusetts Institute of Technology.

Mary P. Anderson is a professor in the Department of Geology and Geophysics at the University of Wisconsin, Madison. Her current research interests include the effects of potential global climate change on groundwater–lake systems and quantifying groundwater recharge. Dr. Anderson received a B.A. degree in geology from the State University of New York at Buffalo and a Ph.D. degree in hydrology from Stanford University. She is a former member of the Water Science and Technology Board.

Roni Avissar is professor and chair of the Department of Environmental Sciences and director of the Center for Environmental Predic-

115

tion at Rutgers University. His research focuses on the study of land–atmosphere interactions from micro to global scales, including the development and use of a variety of atmospheric, land, and oceanic models. Dr. Avissar received his B.S. degree in soil and water science, his M.S. degree in micrometeorology, and his Ph.D. degree in mesoscale meteorology from the Hebrew University in Israel. He is editor of *Journal of Geophysical Research-Climate* and *Physics of the Atmosphere*.

Roger C. Bales is a professor in the Department of Hydrology and Water Resources at the University of Arizona. Dr. Bales conducts research on the hydrology and biogeochemistry of alpine areas, polar snow and ice, and water quality. He received his B.S. degree from Purdue University, his M.S. degree from the University of California, Berkeley, and his Ph.D. degree from the California Institute of Technology.

George M. Hornberger (National Academy of Engineering) is the Ernest H. Ern Professor of Environmental Sciences at the University of Virginia. His current research interests include hydrogeochemical response of small catchments and transport of colloids in porous media. He is chair of the Water Cycle Study Group of the U.S. Global Change Research Program. Dr. Hornberger is a fellow of the American Geophysical Union and a member of the Geological Society of America. He has served on numerous NRC boards and committees, including chairing the Commission on Geosciences, Environment, and Resources. He served as editor of *Water Resources Research* from 1993 to 1997. He obtained his B.S. (1965) and M.S. (1967) degrees in civil engineering from Drexel University and his Ph.D. degree from Stanford University in hydrology in 1970.

William K. Nuttle is an independent consultant in Ottawa, Ontario, Canada. Until recently, he was director of Everglades Department, South Florida Water Management District, and was executive officer for the Florida Bay Science Program immediately prior to that. An expert in ecohydrology of wetlands and environmental science, he has coordinated extensive estuarine and wetlands research programs in South Florida. Currently, he is visiting scholar at the Southeast Environmental Research Center, Florida International University. Previously, he held positions with Memorial University of Newfoundland and the University of Virginia. Dr. Nuttle has also consulted widely on topics generally related to coastal, wetland hydrology and the interface between research and environmental management. He is a member of the Committee on Hydrologic Science. Dr. Nuttle received his M.S. degree and Ph.D. (1986) degree in civil engineering from the Massachusetts Institute of Technology and his BSCE degree from the University of Maryland.

Marc B. Parlange is a professor in the Department of Geography and Environmental Engineering at Johns Hopkins University. His primary research interest is in hydrology and fluid mechanics in the environment, especially questions of land–atmosphere interaction, turbulence and the atmospheric boundary layer, watershed-scale hydrology, and vadose zone transport processes. Dr. Parlange received his B.S. degree from Griffith University (Brisbane, Australia), and his M.S. degree in agricultural engineering, and his Ph.D. degree in civil and environmental engineering from Cornell University.

Kenneth W. Potter is a professor of civil and environmental engineering at the University of Wisconsin, Madison. His teaching and research interests are in hydrology and water resources, including hydrologic modeling, estimation of hydrologic risk, estimation of hydrologic budgets, watershed monitoring and assessment, and hydrologic restoration. Dr. Potter is a past member of the Water Science and Technology Board and has served on many of its committees. He received his B.S. degree in geology from Louisiana State University and his Ph.D. in geography and environmental engineering from Johns Hopkins University.

John O. Roads is director of the Scripps Experimental Climate Prediction Center of the Scripps Institution of Oceanography at the University of California, San Diego. His research focuses on surface water and energy budgets, regional and global climate change, and medium- and long-range weather prediction using modeling and observation. He was a member of the NRC's Global Energy and Water Cycle Experiment (GEWEX) Panel from 1993 to 2000, and most recently chaired the panel. He is a member of the AGU Precipitation Committee. He has been named to advisory panels for the National Climate Data Center and the National Meteorological Center. He received a B.A. degree in physics in 1972 from the University of Colorado and a Ph.D. degree in meteorology in 1977 from the Massachusetts Institute of Technology, Cambridge.

John L. Wilson is professor of hydrology and chair of the Department of Earth and Environmental Science at New Mexico Tech, Socorro. He studies fluid flow and transport in permeable media, using field and laboratory experiments and mathematical models. In the past this has included studies of the movement of water, nonaqueous phase liquids, dissolved chemicals, colloids, and bacteria through porous, fractured, and faulted media. He was the 1992 Darcy Lecturer for the Association of Groundwater Scientists and Engineers. He was elected a Fellow of the American Geophysical Union in 1994. He received the O.E. Meinzer Award from the Geological Society of America in 1996 and was elected Fellow of the Society in the same year. He received his B.S. degree from the Georgia Institute of Technology and his M.S., C.E., and Ph.D. degrees from Massachusetts Institute of Technology.

Eric F. Wood is a professor in the Department of Civil Engineering and Operations Research, Water Resources Program, at Princeton University. His areas of interest include hydroclimatology with an emphasis on land–atmosphere interaction, hydrologic impact of climate change, stochastic hydrology, hydrologic forecasting, and rainfall–runoff modeling. Dr. Wood is an associate editor for *Reviews in Geophysics, Applied Mathematics and Computation: Modeling the Environment*, and *Journal of Forecasting*. He is a member of the Board on Atmospheric Sciences and Climate, the Climate Research Committee, and the Committee on Hydrologic Science. He is a former member of the Water Science and Technology Board and BASC's GEWEX panel. Dr. Wood received an Sc.D. degree in civil engineering from Massachusetts Institute of Technology in 1974.